Basic Engineering Design

Basic Engineering Design

C V Starkey
MPhil., CEng., FIMechE., FIED
Past President of the Institution of Engineering Designers

Edward Arnold
A division of Hodder & Stoughton
LONDON BALTIMORE MELBOURNE AUCKLAND

© 1988 C V Starkey

First published in Great Britain 1988

British Library Cataloguing in Publication Data

Starkey, C. V.
Basic engineering design
1. Engineering. Design
I. Title
620'.00425

ISBN 0-7131-3669-3

All rights reserved. No part of this publication may be reproduced or transmitted in any form or by any means, electronically or mechanically, including photocopying, recording or any information storage or retrieval system, without either prior permission in writing from the publisher or a licence permitting restricted copying. In the United Kingdom such licences are issued by the Copyright Licensing Agency: 33–34 Alfred Place, London WC1E 7DP.

Typeset in 10/11 Times by Mathematical Composition Setters Ltd, 7 Ivy Street, Salisbury, Wiltshire, England, SP1 2AY.
Printed and bound in Great Britain for Edward Arnold, the educational, academic and medical publishing division of Hodder and Stoughton Limited, 41 Bedford Square, London WC1B 3DQ by J. W. Arrowsmith Ltd, Bristol.

Preface

The designer is not only a creator of wealth, but also a creator of expense. So it is very important that the designer understands that every decision brings either a cost benefit to the design, or a cost penalty. In today's competitive world, the economic aspects of design are every bit as important as the technical aspects. A product which fails to meet its cost specification is just as much a failure as one which falls short of its technical specification.

With ever-shortening product introduction lead times and product lives, it is vital for the designer to get it right first time, for there may not be a second opportunity. As more advanced technology becomes available, more and better design decisions are called for. Greater masses of data need to be considered and analysed before those decisions can be finalised. Without help, the designer is in danger of sinking without trace beneath the mass of data.

Fortunately the advancement of technology has thrown out a lifeline for the designer — the microcomputer. It has an enormous appetite for number-crunching, in microseconds and, correctly used, without error. Its memory is vast and instantly on tap. Output can take the forms of visual displays, hard copy, graphics, 3-D models, direct instructions to machining centres, and many others. It can help the designer with discretionary decision-making, by replacing intuitive techniques with exhaustive analyses from huge databases in frighteningly short times. In short, the microcomputer can help the designer to make more effective use of his or her time, and can improve the level of the cost-beneficial decisions.

The language common to most microcomputers is BASIC — an anagram for Beginners All-purpose Symbolic Instruction Code. This work attempts to combine the language of BASIC with the elements of engineering design. The programs herein are in the 'Microsoft' version of BASIC and they are elementary enough for ready conversion by the student to other dialects of BASIC. The programs within the text are essentially simple. Each explores one aspect only in order to keep the level of comprehension within the scope of the student.

The second part of this work contains a design assignment which takes the reader through the evolution of the design of a product, in the light engineering field, from initial requirements to final general proposals.

Design is essentially a learn-by-doing activity. It is hoped that this guided design assignment will provide a well documented model of actual design procedure for future reference and use by the reader. It is particularly aimed at developing in the student the 'techne' qualities required by the post-Finniston CNAA degree of B. Eng. It is also apposite for students of TEC HC and HD courses in engineering design at levels 4 and 5. In this section of the work, it has sometimes been necessary to extend the BASIC programs to embrace more than just a single aspect. In these instances, data sheets have been included which show the sequential progression of the calculations, which should simplify the understanding of the resultant programs.

Finally it must be stressed that this work makes no attempt to teach technology or computer programming. It shows only how technological and computational aspects are utilised in the design function. It is expected that readers will be competent in their knowledge of materials science, thermofluid mechanics, electrotechnology, dynamics of systems, computer studies, etc, from other parallel work. Some references to suitable reading in appropriate subjects are given in the bibliography.

Material based on PD6470: 1981 is reproduced by permission of BSI. Complete copies can be obtained from them at Linford Wood, Milton Keynes, MK14 6LE.

Clifford Victor Starkey
1987

Contents

Part 1

THE CRAFT OF DESIGN

1 Introduction

So what is Basic engineering design? We might start by asking the question, 'What is engineering design?
It is

> the recognition and understanding of a basic need and and the creation of a system to satisfy that need.

Put more simply, design is problem finding followed by problem solving.
Clearly, a poor solution to the correct problem is infinitely more useful than an elegant solution to the wrong problem. And since it is pointless to solve the wrong problem, it is self-evident that

> **problem finding is more important than problem solving.**

Einstein had a few words to say on this subject:

> The formulation of a problem is often more essential than its solution, which may be merely a matter of mathematics or of experimental skill. To raise new questions, new possibilities, to regard old problems from new angles requires creative imagination and marks real advance in science.

It is often claimed that it is impossible to teach design, and that designers are born not made. Certainly, aptitude and flair are useful qualities in a designer but, however, naturally talented he or she might be, performance can always be improved by professional training. You would not choose to have your teeth fixed by a 'born' dentist; and you must never trust your product design to a 'born' designer.

Design method

So, if we accept that the teaching of design is possible, we must also accept that it is possible to apply some sort of logical method to that teaching. Such a logical method will inevitably consist of a number of sequential steps.
Although there is much argument about how many steps there should be (see references 1, 2, 10 and 14), and also what detail each step should contain, the following seven-step method is recommended as being adequate to deal with most design problems.

A PROBLEM FINDING
1 Identify the basic need
2 Define the problem to be solved, arising from that need
3 State the parameters within which the solution must fit

B PROBLEM SOLVING
4 Create lots of ideas which might form suitable solutions
5 Evaluate each of these ideas
6 Isolate the idea giving the best solution to the problem
7 Implement the solution to the problem.

Let us pause for a moment to review the method which is set out above.

Problem finding requires the designer to exercise 'detective' qualities of objectivity, analysis and synthesis. He or she must put aside subjective and emotive feelings and believe only those statements which can be 'proved' or, at least, can be substantiated from more than one source.

Problem solving requires the designer to discard the 'detective' qualities and give full rein to more creative abilities. Emotion, subjectiveness and even frivolity must be freely expressed. This is the complete antithesis of the problem finding activity.

It is clear from the method propounded that there is no unique solution to any design problem. There may be several hundred potential solutions of which perhaps ten or fifteen may be quite elegant. It is then up to the designer to decide which of the alternative solutions is the best. This is quite unlike the usual numerate problems posed by engineering science where the allocation of values to the variables in an equation will yield a single, correct answer. **This is perhaps the major difference between the work of the designer as compared with that of an engineer.**

In the above design method, the relative importance of each of the seven steps is less than that which it follows.

Introducing Pareto

Vilfredo Damaso Frederico Pareto (1848–1923) was an Italian sociologist. Born and brought up in France where his father, the Marquis Raffaele Pareto, a member of Mazzini's party, had been obliged to emigrate because of his revolutionary activities, Vilfredo eventually became Professor of Political Economy at Lausanne in Switzerland. He published a number of standard works on economics and perhaps his best known work is the Pareto distribution. This relates the number of elements in a whole to the effect which those elements exert on the whole.

For example, he examined the distribution of incomes in a typical community and discovered that

75% of community income was earned by 20% of inhabitants
18% of community income was earned by 30% of inhabitants
7% of community income was earned by 50% of inhabitants

Plotted graphically, these figures produce a typical Pareto distribution with its distinctive **knee** configuration, as shown in Figure 1.1.

Pareto distributions may be found in any area where a number of elements exert influence on a whole activity. For example, the value of items in the weekly shopping basket, the value of components in a factory storeroom, the creation of wealth by industry and commerce (the top 200 companies), active membership of building societies, active participants in trades unions activities, etc.

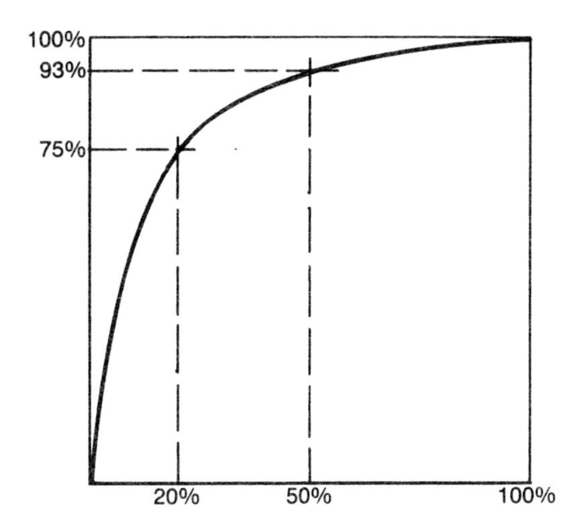

Figure 1.1 Typical Pareto distribution

But what interest has Pareto for the designer? Let us return to the seven-step design method. We may say **the whole of the design method is composed of seven elements in descending order of importance** and so the idea of a Pareto distribution between these elements and their effects becomes irresistible.

This can be set out in tabular form as in Table 1.1. It must be clearly understood that the values assigned to **effect** are notional only.

The seven-element design method can also be represented by a logic diagram as shown in Figure 1.2.

It should be noted that the element of **evaluation** is achieved by comparing each created idea with all the practical constraints and with the parameters within which the eventual solution must fit. Some of these constraints will be external to the design problem; for example they may be legal, technical, social, financial, ecological. Others will be internal to the design problem and will arise as a result of conflicting requirements within the design itself. Only when a compatible interface exists between an idea and all these practical constraints, can that idea become a candidate for final acceptance as a solution.

The process shown in the logic diagram is iterative, beginning with the broad concept

Table 1.1 Pareto relationship of elements to whole design function

Element	Effect	Phase	Function
Identification	0.50	Recognition and	
Definition	0.20	understanding	0.80 Problem finding
Parameters	0.10	of basic need	
Creation	0.08	Creation of a	
Evaluation	0.06	system to	0.20 Problem solving
Isolation	0.04	satisfy that	
Implementation	0.02	need	

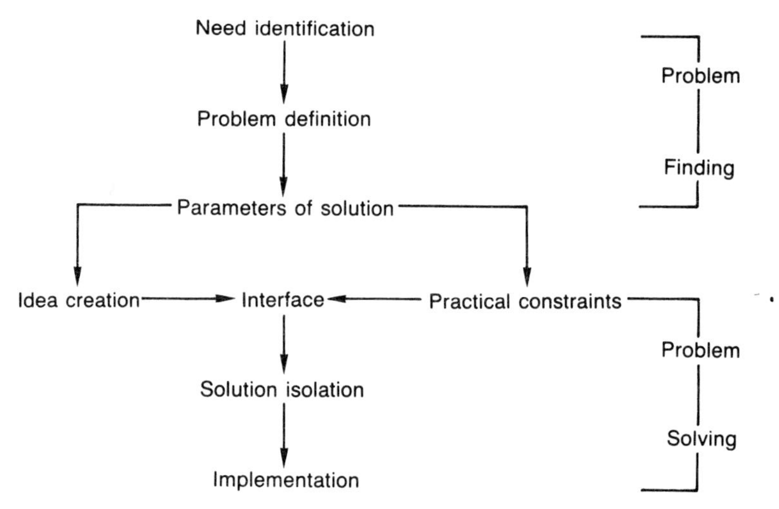

Figure 1.2 Logical progression of the design function

of a general solution and then progressing through intermediate levels to the final solutions of a myriad detailed design problems. As individual problems at these various levels are identified and solved, it is entirely possible that some of the information generated may modify the original specification as outlined in elements 2 and 3, or it may become part of the practical constraints acting upon the design solution.

We have already defined design as **the recognition and understanding of a basic need and the creation of a system to satisfy that need.** The use of the word **system** is justified in that the solution to a design problem may not necessarily result in the production of an item of hardware. For example, the creation of a railway timetable is a design problem which results in the production of an item of software.

Looking in greater detail at what is actually involved in the seven-element design method, we may say that the designer should be trying to achieve these general objectives.

seeking information as to what the basic need is
understanding how the basic need arises
establishing the environment within which the need must be satisfied
manipulating resources to ensure optimum satisfaction of the need
ensuring that the general environment is not degraded by the solution.

Need identification

Who needs it?
How does the need manifest itself?
Is it recognised by sight, sound, smell, taste or touch?
Is it recognised by a combination of the above?
Is it abstract?
What is the environment within which it (the need) exists?
What resources, if any, does it now consume?
When is it needed and for how long?

Where is it needed?
Why cannot it be satisfied by an existing system?

These and many similar questions must be posed to determine what the actual need is. They do not attempt to postulate what the solution might be. It is essential that the designer resists any inclination to jump to conclusions at this stage. The success of all the work which will follow this element depends upon the thoroughness with which the designer pursues the true identity of the basic need.

Unless the need is properly identified, there will be little success in finding an acceptable solution, and the real basic need will not be fully satisfied. In searching for that real basic need, the designer must be prepared to disbelieve anything he or she is told, unless it can be verified by alternative means. Questioning must be deep and penetrating and the designer must seek practical demonstrations of the need from those who have it.

Where the need is exhibited by only one person or one group, this may be easy. For example, the design of a walking aid for arthritis sufferers may involve the close questioning of many patients in order to determine how the basic need may be affected by different levels of disability. But, clearly, there are fairly close limits controlling the total extent of the need which will be revealed.

On the other hand, if the designer is concerned with the design of a car, it will not be possible to question every potential customer. Also there may be other interests which have to be satisfied within the basic need. These may include those of the manufacturer, the distributors and the garages, in both home and foreign markets. In these circumstances the search for the basic need is complex in the extreme, and the designer will be forced by expedient to accept secondhand opinion and to act upon it. It is, therefore, essential to be thoroughly satisfied as to the integrity of the information sources and the validity of their data.

Problem definition

In defining any problem, words are the designer's worst enemy. Too much description will confuse and conceal the true meaning of thoughts. Brevity is essential. Designers must try to confine themselves to two words only, a verb and a noun. In this way the essence of the problem is preserved, uncluttered by verbiage. To demonstrate, we might consider the prime functions of a number of everyday items and describe them using only a verb and a noun.

Item	Prime function
teacup	contain fluid
kettle	heat fluid
chair	support mass
sack barrow	transport freight
washing machine	launder clothing
floor cleaner	clean floor
ball pen	make marks
spectacles	correct sight
ruler	measure distance
wrist watch	display time

Here the verb/noun combination is used to describe the primary function of the item. But most items also have a number of secondary functions to perform.

The aesthetics of a wine bottle were once described thus.

> The base is ruggedly functional, promoting an impression of great stability. At the other end, the neck is gently sculptured to assist the precision transfer of the vessel's contents into small wineglasses. The sides connecting these two features carry through the design theme by their transition from the robust base to the delicate neck — *and are also useful for sticking the labels on.*

Although mickey-taking, it does underline the multifunctional roles of even the simplest component. And it is very easy to overlook some of these secondary functions when designing items.

Let us look at some typical secondary functions for some of those items listed above.

Item	Prime function	Secondary functions
teacup	contain fluid	insulate heat, be decorative, be inexpensive
washing machine	launder clothing	dry clothing, look good, be quiet, be safe
kettle	heat fluid	insulate heat, pour fluid, look good, be reliable

So although sometimes only two words seem insufficient to convey the full definition of the item, usually they will suffice but occasionally a third word might be added to clarify meaning.

Parameters of solution

Any problem may be solved given enough time and enough money. But most real problems have to be solved in real time and at minimum cost. Realistically, budgets must be preset for all the resources that are likely to be consumed in solving the problem, and their consumption must be strictly monitored. Only five resources are available for manufacturing; they are

labour, materials, time, space, money

and limits on their usage must be stated before the problem solving phase begins. Let us define these five resources in a little more detail.

labour	skill requirements, technology levels
materials	manufacturing materials, energy sources, equipment
time	timescale for implementing the solution
space	location where the solution will be implemented
money	cost of solution implementation

Important! During the problem finding phase great care must be taken to preserve total objectivity. We are not yet at the decision making stage and on no account must any attempt be made to anticipate the form of the final solution. Indeed, there is considerable advantage in delaying the start of the decision making process for as long as is reasonably possible, as this gives extra time for the mental assimilation of the basic need and the design problem arising therefrom. There is overwhelming evidence that full mental absorption of the design problem requirements, followed by an incubation period of subconscious analysis and synthesis, often leads to that elusive inspirational flash which can set the designer on the track of a successful solution.

Idea creation

This is the first of the creative elements of the design process, and is perhaps the most difficult. The object is to promote the free flow of ideas, any or all of which may be candidates for adoption as the final solution.

The designer should not be restricted to one central idea but should give free rein to the imagination and allow the ideas to flow without interruption. **On no account must he or she be tempted to judge the individual ideas as they come.** Write them all down; evaluation comes later. Inevitably the flow of ideas will eventually dry up, and more will be said later about how the flow can be restarted.

The result from this element should be a number of raw solutions which roughly satisfy the parameters already set out, before the unavoidable concessions and compromises which will be necessary to make any solution totally acceptable in its technical, legal, social and ecological requirements.

Idea evaluation

In this element the designer progressively examines each idea that has been generated against the parameters of an acceptable solution which has been set out. Each examination is a small feasibility study aimed at either eliminating the idea or passing it on for further consideration.

The designer must ensure that each possible solution is both externally and internally compatible. For example, any solution which had as a by-product a noxious effluent, whether liquid, gas, noise or other irritant, could not be regarded as externally compatible with social and ecological parameters, unless the designer could complete the cycle to neutralise such effluent.

Similarly, most designs are of multi-component form, and each individual component must be internally compatible with those adjacent to it in the final assembly. For instance, two components, separately designed, which ultimately must be joined together may be of dissimilar materials resulting in corrosion at their interface due to electrolytic action.

Once some design decisions have been made, the designer may find the freedom for decision making in associated areas severely limited, so that compatibility checks are essential. Generally, the examination will consider firstly, technical aspects of labour, materials, time, space, money; then secondly, social, legal and ecological implications.

Solution isolation

The final choice of the idea which is most appropriate for the actual solution to the design problem may already be obvious through the screening process of evaluation.

However, if several ideas have survived thus far, and all appear to be equally attractive as the final solution, the ultimate choice may be made by selecting that idea which has the lowest degree of difficulty in its implementation.

Implementation

This is the final element in the design process. It includes the commissioning of the design details to the media necessary for system manufacture, be these drawings, magnetic tapes, or any other form of communication which may be appropriate. This element also includes the issue of all manufacturing and testing data, and any information which may be necessary for the marketing, distributing and servicing of the design.

Simple design exercise

To complete this section, let us consider a very simple design problem and see the effect of applying to it the seven-element design method.

Figure 1.3 shows a sketch of a travelling electric razor set. The case comprises two plastic mouldings hinged together to form base and lid. The base moulding has an integral internal fin to separate the razor and the mains lead and cleaning brush. Cemented into the lid moulding is a glass mirror, presumably to assist the shaving

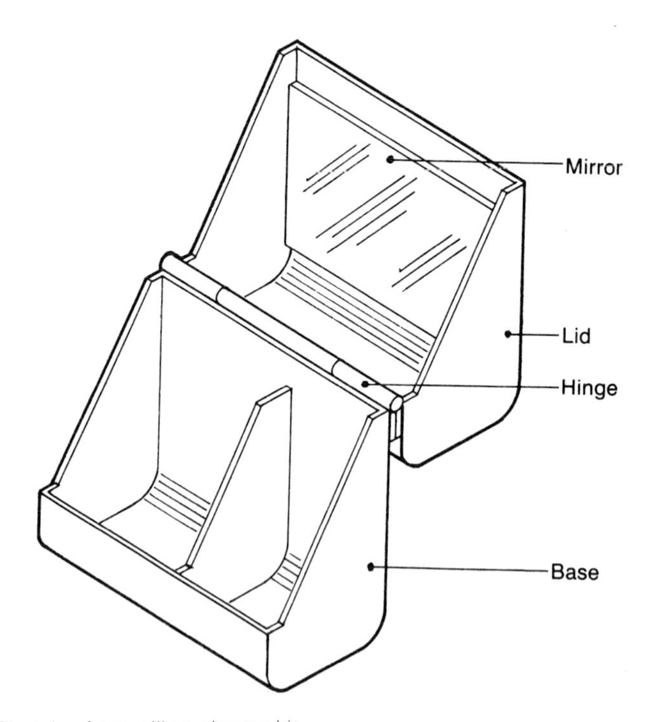

Figure 1.3 Sketch of travelling shaver kit

operation. When the shaver and accessories are removed and the case is stood on a flat horizontal surface, the case overturns due to the offset mass of the mirror and lid. In order to be able to shave with both hands free and unencumbered by having to hold the case and mirror with one hand, and to be able to see one's face, we need to achieve a stable, vertical mirror. And we need to do this in a short time and at minimum cost.

So let us use the seven-element design method to try for a reasonable solution to the problem.

Identify need
    To see face while both hands are free for shaving

Identify problem
    Stable mirror (verb/noun)

Parameters of solution

labour	myself
materials	DIY and general household
time	say, a couple of hours
space	any temporary or permanent quarters I may occupy
money	maximum 10% of shaver price, say £3.00

This completes the problem finding part of the exercise and gives a clear indication of precisely what limitations exist in formulating a reasonable solution to the present problem.

Ideas creation

Use existing case
- prop case
- counterbalance case
- hang case
- turn case
- stick case

Modify existing case
- detach mirror
- hinged supports
- separate case-halves
- extend base

Modify method of use
- separate mirror
- barber shave
- shave by touch
- closed-circuit television
- stop shaving

Before evaluating the very many ideas which can be derived from the fourteen specimens above, we need to ask three questions ...

is the problem real or imagined?
are we using the present facilities correctly?
do we need to do anything?

If the answers to these three questions are — **real, yes, yes** — then we can proceed with evaluation.

Evaluation

Idea	Lab	Mat	Time	Space	Money	Social	Legal	Ecol	Comment
prop case	x	x	x	x	zero	x	x	x	use book to prop case
c/balance	x	x	x	x	£3.00	x	x	x	may hamper portability of razor kit
hang case	x	x	x	x	zero	x	x	x	over edge of windowsill or shelf
turn case	x	x	x	x	zero	x	x	x	turn case on its side
stick case	x	x	x	x	10p	x	x	x	Blutak under base
det mirror	x	x	x	x	zero	x	x	x	may get lost or broken
hinge sppt	x	x	x	x	50p	x	x	x	card stuck to back of case
sep halves	x	x	x	x	50p	x	x	x	hinge pin may get lost
extend base	x	x	x	x	£1.50	x	x	x	may hamper portability of razor kit
sep mirror	x	x	x	x	£2.00	x	x	x	may get lost or broken
barber	?	?	x	?	£2/day	x	x	x	too expensive
by touch	x	x	x	x	zero	x	x	x	shave not very good
cctv	?	?	?	?	£1000	x	x	x	too expensive
stop	x	x	x	x	zero	x	x	x	do I want a beard?

Note that even fairly ridiculous ideas like cctv and stop shaving have been included, because they could be real solutions given the right conditions, and also because there must be no attempt to judge ideas before the evaluation element, as this might stop the flow of ideas.

Solution isolation

Idea	Degree of difficulty 0 to 10
prop case	2
hang case	0
turn case	0
detach mirror	5
by touch	3
stop shaving	0

Choice must be made from three possibilities. Assuming one does not wish to grow a beard, choice is reduced to two, both of which are equally acceptable and either of which may be used at will.

Before passing on, just a few comments on the above exercise. The problem finding part is fairly straightforward, although a note to concentrate on the mirror rather than on the case might direct the attention more positively to the real problem. In the case of ideas creation, the various ideas generated have almost totally been described by the two word method, but the designer should feel free to slip in an extra word if this will clarify the idea. During evaluation all ideas have been looked at regardless of whether they are serious contenders. In this way other, related, ideas may be generated, rather as in a value engineering brainstorming session. In the listing of ideas a cross indicates compatibility while a query indicates incompatibility.

Helping ideas flow

It was observed earlier that eventually the flow of ideas will dry up and must be restarted. We shall now consider some of the ways in which the flow can be got moving again. As has been previously stated, **the essence of the design process is that there is no unique solution to the design problem**, so the more ideas that can be generated, the better the chance of achieving a good solution.

1 Function listing

During problem definition, the idea was explored of using a verb and a noun to describe the primary and secondary functions of everyday items of hardware. By listing and examining all the primary and secondary functions of an article, we can help restart the flow of ideas.

For example, consider the design of a domestic floor cleaner. Its primary function is to **clean floor** but it also has a number of secondary functions, such as **store dust, illuminate floor**, **be safe**, **look good**, **be quiet**, **be inexpensive**, etc. We should examine each of these functions and suggest ways in which they might be achieved.

clean floor	suck, blow, beat, sweep, wash, mop
store dust	plastic container, fabric bag, paper sack, dust pan
illuminate	candle, torch, gas jet, headlamp, room light
be safe	low weight, smooth contours, double insulation
look good	shape, colour, size, texture
be quiet	noise insulation, manual operation, ear plugs
be inexpensive	income range, competition, value for money

If each of these possible methods of achieving a particular function is looked at in detail, ideas will begin to flow again adding fresh material for consideration.

2 Input–control–output

Any piece of hardware which has a working function can be described as **converting available inputs into required outputs through system control**.

available inputs → system control → required outputs

Staying with the floor cleaner example, and assuming we decide to design it as a vacuum cleaner, we can specify the available inputs and the outputs which are required.

Inputs	intelligence, muscle power, 250 V 50 Hz
Outputs	suction, beating, sweeping, storage, illumination

It should also be possible to trace through the path connecting each input with each output, as in Figure 1.4.

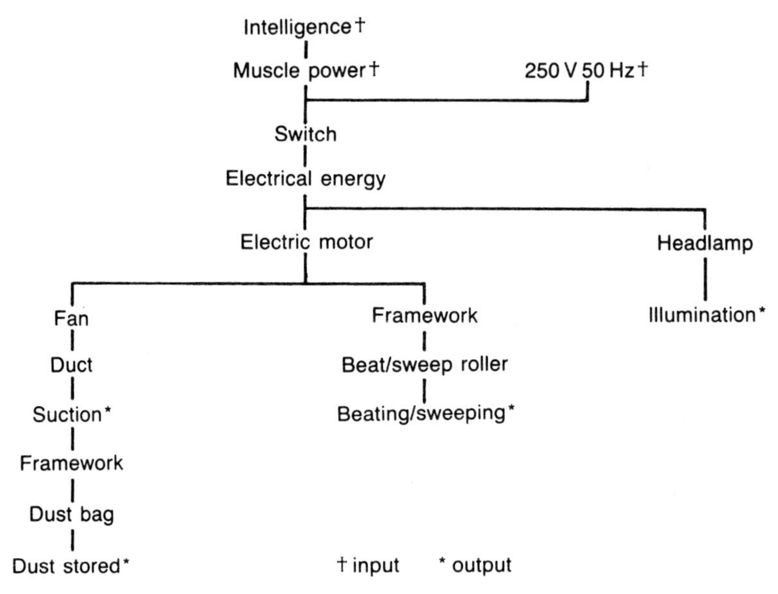

Figure 1.4 Relationships of inputs to outputs

A detailed examination of the various flow diagrams thus produced will again stimulate the flow of ideas, by posing a number of questions. For instance, consider the electric motor ...

what sort of electric motor is needed?
how much power must it develop?
what are its torque and speed requirements?
how will I build it into the framework of the cleaner?
how will I get electrical power to it?
where will I position the on/off switch for controlling it?
what safety devices must I provide?
how will all these items affect the cleaner's appearance?
and so on, and so on ...

For each of these questions a satisfactory answer is required and, as has already been said, there will be no unique answer to any of them. The designer must decide which answers will give the optimum results. More on how these decisions are made later.

3 Attribute listing

In stating the primary and secondary functions of the floor cleaner we were, in fact, setting out the features which the end-user might expect to find in the finished product. For example ...

clean floor, store dust, illuminate floor, are all measures of performance
look good, is a measure of appearance
be safe, be quiet, are measures of safety
be inexpensive, is a measure of cost

These features, performance, appearance, safety and cost are called **attributes** of the finished design. There are very many such attributes which may be desirable in the finished product. Just a few of the more popular ones are listed below, together with some of the factors to be considered by the designer when selecting a particular attribute.

Attribute	Factors to be considered
appearance	shape, size, colour, texture
capacity	size, force, movement, direction
cost	initial, running, maintenance, trade-in
development (of)	performance, reliability, safety, weight
durability	abuse, misuse, accident, corrosion, humidity
interchangeability	rapidity, accuracy, multi-role, modules
life	first line, backup
maintainability	continuous, regular, sporadic, none
performance	force, velocity, pressure, energy
portability	lift, transport, orientation
reliability	MTTF, MTBF*, repeatability
safety	electrical, mechanical, chemical, noise
simplicity	technology, manufacture, maintenance, use
serviceability	in situ, remote
weight	amount, distribution

*MTTF = mean time to (first) failure
MTBF = mean time between failures (see reference 9 for details)

Returning to the floor cleaner, we can express all its primary and secondary functions in terms of attributes.

performance	clean floor, illuminate floor, store dust
appearance	look good
safety	be safe, be quiet
cost	be inexpensive

What we are saying is that the design of the floor cleaner can be described in design terms as consisting of

performance, appearance, safety and cost attributes.

In real life, however, the end-user would also be interested in some of the other attributes which we have listed. Attributes such as life, reliability, durability, and perhaps many more. All these attributes can be used as a guide to the creation of a 'perfect' floor cleaner, aimed at satisfying the greatest possible number of potential users. However, the more attributes required, the greater the difficulty for the designer to satisfy all of them.

Principal attributes

In any selection of desirable attributes used to describe the features of a product, the following five will usually appear somewhere in the top ten listing,

appearance, cost, performance, reliability, safety.

The five, which are given here in alphabetical order so that none will be favoured more than any other, may be regarded as **principal attributes**, appearing most frequently and being of first line importance. To recap, consider how these five principal attributes can be used to describe the requirements of the floor cleaner.

appearance	shape, size, colour, texture
cost	initial, running, maintenance, trade-in
performance	force, velocity, pressure, energy
reliability	MTTF, MTBF, repeatability
safety	electrical, mechanical, chemical, noise

By expanding these five principal attributes to an appropriate level of detail, one can completely describe all the essential features of the floor cleaner. This will provide the designer with a target against which to measure his or her efforts.

As the design proceeds, it is necessary for the designer to make some **trade-offs**. Perhaps a bit more performance can be squeezed from the product, but only by making it more expensive. Or maybe the safety of the product can be improved, but only by detracting from its appearance. Which way should the decisions go?

It would be helpful if the principal attributes could be given some priority, the most important at the top of the list and the least important at the bottom.

Attribute priority

Clearly, the importance of an individual attribute will depend largely on the product in which it features. And we must expect different products to contain the principal attributes in different priority orders.

In an industrial product, such as a machine tool, performance and reliability may be of prime importance with cost, safety and appearance being less critical (see Figure 1.5).

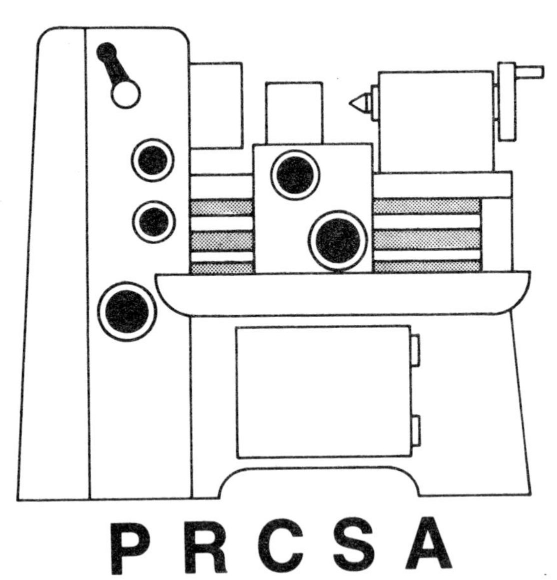

Figure 1.5

On the other hand, a domestic product, such as a washing machine or a vacuum cleaner, might give priority to appearance and cost with safety, performance and reliability in supporting roles (Figure 1.6).

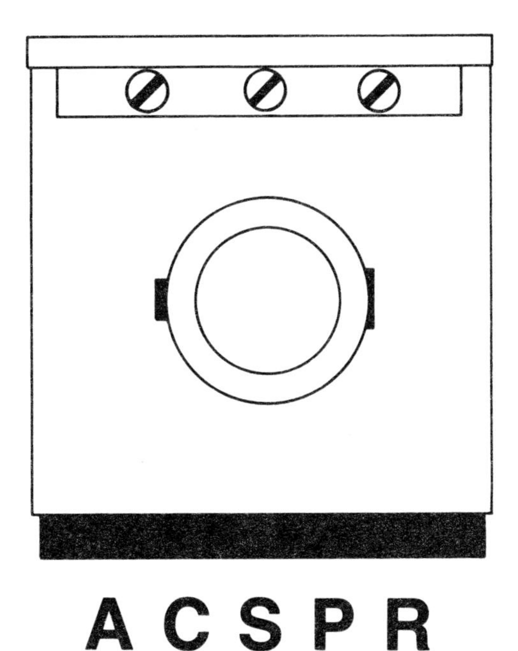

Figure 1.6

A utilitarian product, such as an invalid aid, might require safety and reliability to be at the top of the list and relegate performance, cost and appearance to lower levels of importance (Figure 1.7).

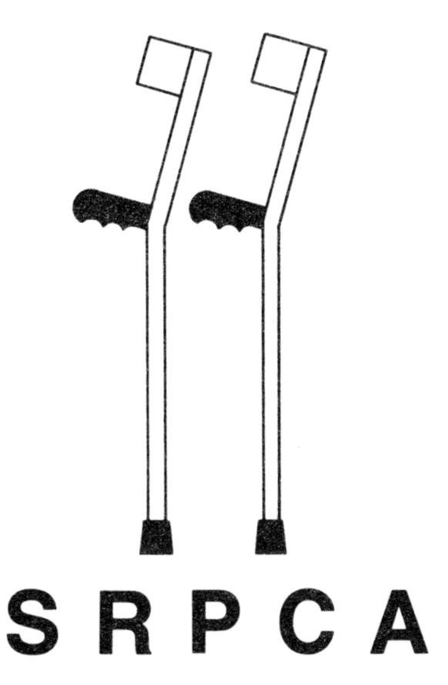

Figure 1.7

In a 'luxury' product, such as a childs toy, appearance and performance might be the prime influences of choice with safety, cost and reliability being relatively unimportant (Figure 1.8).

Figure 1.8

In the case of a better mousetrap, perhaps the order should be as shown below... strictly an Anglo-Saxon choice (Figure 1.9).

Figure 1.9

It must be emphasised that such choices of priority are largely notional. One can never say that safety in a childs toy is *unimportant*; merely that it may be less important in influence with the purchaser than appearance or performance.

Such priority listings of principal attributes are subjective. They can be made somewhat more objective by comparing them in pairs when making choices; this enables concentration on just the two under consideration, to the exclusion of all others. To demonstrate, consider how one might decide which of three alternative cars to purchase.

We list the three cars and then compare them, two at a time, awarding a mark to the preferred vehicle of the pair and zero to the other.

	Comparisons				
Car	A/B	A/C	B/C	Total	Priority
A	1	1		2	1
B	0		0	0	3
C		0	1	1	2

In column A/B we compare cars A and B, decide that we believe A is superior to B and award it one mark, giving zero to B.

In column A/C we compare cars A and C, decide A is superior to C and award it 1, giving 0 to C.

We now know that we prefer car A to both cars B and C and, given a free hand, we would go out and buy car A.

But suppose car A is not available for some reason. Perhaps it is too expensive, or perhaps it is not available in the colour we want, etc. In this case we may be forced to buy our second choice. But which is that, B or C? Before we can decide we must compare B with C and choose the superior model. This we do in column B/C, and we may decide in favour of car C.

The 'total' column sums the marks awarded to each vehicle and from this we can deduce the priority order.

The total number of decisions necessary to establish priority is ...

$D = n(n-1)/2$ where D is the total number of decisions
and n is the number of alternatives.

This technique can be applied to establish priority order for the five principal attributes in any product, in which case the total number of decisions necessary would be

$$5 \times 4/2 = 10$$

However, such a priority listing can indicate only the **pecking order** of importance of the five principal attributes. It would not indicate the relative importance between attributes, and a knowledge of such relationships would be of great help to the designer in his trade-off decision making. What is needed is some sort of relative rating for the principal attributes.

Attribute rating

On page 17 is shown a priority listing of principal attributes for a domestic product such as a vacuum cleaner. It is

appearance
cost
safety
performance
reliability

and this we believe to be an **attribute model** of the ideal vacuum cleaner, as seen through the eyes of the end-user.

When deciding which of several alternative vacuum cleaners to buy, the attributes of appearance and cost are immediately evident to the consumer, and he or she can make direct comparisons between alternative models in terms of both appearance and cost. This is exactly similar to our hypothetical comparison of three cars.

The attribute of safety is not so evident at the point of purchase, although a rough idea of the safety features of each product can be gained by looking at such items as electrical insulation, mechanical guarding, duty rating of electrical components, etc.

As far as performance and reliability are concerned, the prospective purchaser can have no firm information other than the claims made by the various manufacturers. Unless the user can have some lengthy experience of using every model of cleaner no real knowledge of these two attributes can be acquired.

Thus the decision as to which cleaner to buy is largely made on the visual evidence of appearance and stated price, perhaps influenced by any secondhand advice from friends or relatives, plus some rather vague idea of the relative safety features of the models under consideration. One of the prime jobs of the product designer is to try to anticipate the reactions of the potential customer and so influence the choice of purchase.

When the designer begins the design of a product, a priority listing of the principal attributes is very helpful. Further, a knowledge of the relative importance of each attribute in the listing would be an invaluable aid in decision making. Clearly, if the relative difference between cost and performance is only 10%, then any trade-off decision involving these two attributes must be made much more carefully than if the difference is, say, 80%.

A relative importance rating of attributes can be arrived at in a number of ways.

1 We can just go ahead and assign values intuitively, for example

0.35 appearance
0.30 cost
0.15 safety
0.12 performance
0.08 reliability
1.00 total

2 Alternatively, as we have a number of elements of differing values making up a whole, we can apply a Pareto distribution thus...

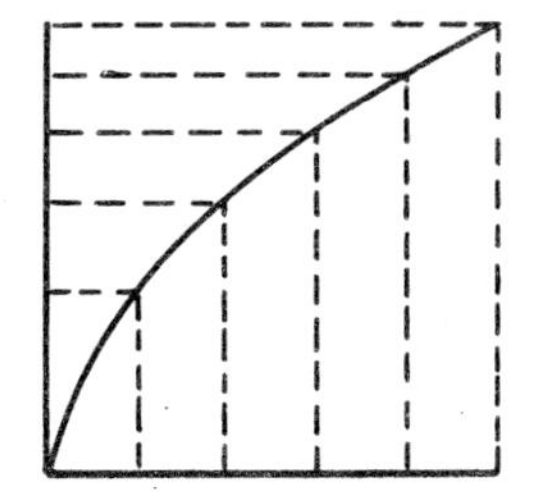

0.34 appearance
0.24 cost
0.18 safety
0.13 performance
0.11 reliability

3 As suggested by Svensson (14), we could adopt a value scale of importance which approximates to a linear scale, and the set (1, 3, 5, 8, 10) is considered suitable.

10 appearance
8 cost
5 safety
3 performance
1 reliability
27

For direct comparison with the values shown in items 1 and 2 above, we need to normalise this listing to unity total. We do this by dividing each value by the total value, thus ...

10 appearance	becomes 0.370
8 cost	becomes 0.296
5 safety	becomes 0.185
3 performance	becomes 0.111
1 reliability	becomes 0.037
27	becomes 1.000

4 Finally, we could build a rating procedure into our method of determining the original priority listing, thus killing two birds with one stone. Let us return to the pairs–comparison method, and instead of allocating one and zero to our choices, we mark the two attributes under consideration from a total of 100 points. Start by listing the attributes in alphabetical order.

	Comparisons			
Attribute	1/2	1/3	1/4	1/5
1 appearance	55	70	80	65
2 cost	45			
3 performance		30		
4 reliability			20	
5 safety				35

then compare attributes in pairs, 1 with 2, 1 with 3, etc., each time allocating a value to each attribute, representing your value judgement of their individual worth, so that the two values when added equal 100. In this case the total number of decisions is

$D = n - 1$ where D is the total number of decisions
and n is the number of alternatives

On page 00 it was suggested that 10 total decisions would be necessary to assign priority order to five attributes, i.e. $D = n(n-1)/2$, but now it is suggested that only four total decisions are necessary, i.e. $D = n - 1$. The reason for the reduced number of total decisions is, that by assigning individual values to a total of 100 points for each attribute comparison, we have automatically set the relative values for all other subsequent choices.

For example, the relationship between, say, performance and cost can be shown to be

performance/cost = performance/appearance × appearance/cost
in column 1/2, appearance/cost = 55/45
in column 1/3, performance/appearance = 30/70
so that performance/cost = 30/70 × 55/45 = 0.429 × 1.222 = 0.524

From this we see that performance = 0.524 × cost
but we also know that performance + cost = 100 (total allocated)
so it is clear that cost (1 + 0.524) = 100
and cost = 100/1.524 = 65.62
and performance = 100 − 65.62 = 34.38

In this way we can express the relative value of all the attributes as a factor of one of them. Take the first attribute of appearance as the base for comparison.

Start with the attribute of safety which has a value of 35 points. Its relationship with appearance is 35/65, a factor of 0.538.

Next look at reliability with a value of 20 points. Relative to appearance it has a factor of 20/80 = 0.250.

The factor for performance is 30/70 = 0.428.

The factor for cost is 45/55 = 0.818.

Finally, the factor for appearance is 55/55 = 1.000.

The completed pairs-comparison now looks like this.

	Comparisons				
Attribute	1/2	1/3	1/4	1/5	Factor
1 appearance	55	70	80	65	1.000
2 cost	45				0.818
3 performance		30			0.428
4 reliability			20		0.250
5 safety				35	0.538

The priority order, **with ratings**, is

1.000 appearance	normalised to 0.33 appearance
0.818 cost	normalised to 0.27 cost
0.538 safety	normalised to 0.18 safety
0.418 performance	normalised to 0.14 performance
0.250 reliability	normalised to 0.08 reliability
3.034 total	normalised to 1.00 total

Over the last few pages we have been **making decisions**. Before we go on to explore decision making in detail, let us see if we can consolidate the present situation. The analysis of attributes and the establishment of priority and ratings is essential for good decision making and hence good design. But the involvement of so much arithmetic, albeit simple in content, is a chore. So let us hive off to the microcomputer all the arithmetic, for which it is ideally suited. Let us look at very simple computer aided decision making.

Computer aided decision making (CADEMAK)

At any level, decision making is essentially a matter of choosing between a number of alternatives. Such choices are totally subjective, being based on our knowledge of the alternatives available and on the sum total of our personal experience and culture.

Choice suggests numerical weighting. We decide that one course of action is better than some other course, by some arbitrary amount, '*this alternative is ten times better than that*'. Using numerical values to quantify the preferred choice reduces intuitive decision making to arithmetical calculation. This is decision making at its crudest; if increased refinement is required, other factors can be built into the equations. Looking again at our attribute model, the essential procedure is

decide the principal attributes, how many and which ones
decide relationships between them by pairs comparison
compute factors relating importance of all attributes
normalise these factors to unity total
arrange attributes in descending order of importance
print the model–priority listing with ratings

In computer language this is

data input
computation
bubble sort
output

Add a title and we have a useful decision making computer program.

In Figure 1.10

lines 100–150 title of the program
lines 160–240 data input, principal attributes and primary decisions
lines 250–440 compute factors relating attributes in two-dimensional array

```
5 PRINT CLS
100 PRINT"COMPUTER AIDED DECISION MAKING
110 PRINT"______________________________     ________
120 PRINT:PRINT
130 PRINT"PHASE 1    SETTING UP THE MODEL
140 PRINT"______________________________
150 PRINT
160 INPUT"HOW MANY PRINCIPAL ATTRIBITES ";N
170 PRINT
180 PRINT"GIVE PRINCIPAL ATTRIBUTES ALPHABETICALLY
190 PRINT
200 DIMA$(N)
210 FORJ=1TON:INPUTA$(J):NEXTJ
220 PRINT
230 PRINT"GIVE PRIMARY DECISIONS
240 PRINT
250 DIMM(N,N-1)
260 FORJ=1TON-1
270 INPUTM(1,J)
280 NEXTJ
290 PRINT
300 DIMR(N)
310 LETR(1)=1
320 FORJ=1TON-1
330 K=J+1
340 LETM(J+1,J)=100-M(1,J)
350 LETR(K)=M(J+1,J)/M(1,J)
360 NEXTJ
370 FORJ=1TON
380 LETB=B+R(J)
390 NEXTJ
400 DIMF(N)
410 FORJ=1TON
420 LETF(J)=INT((R(J)/B)*100+.5)/100
430 NEXTJ
440 PRINT
450 PRINT
460 PRINT"MODEL SOLUTION
470 PRINT"______________
480 PRINT
490 FORI=1TON-1
500 FORJ=1TON-1
510 IFF(J)>=F(J+1)THEN580
520 LETT=F(J)
530 LETV$=A$(J)
540 LETF(J)=F(J+1)
550 LETA$(J)=A$(J+1)
560 LETF(J+1)=T
570 LETA$(J+1)=V$
580 NEXTJ,I
590 FORI=1TON
600 PRINTF(I),A$(I)
610 NEXTI
620 END
READY.
```

Figure 1.10 Computer aided decision making program (CADEMAK)

```
COMPUTER AIDED DECISION MAKING
______________________________

PHASE 1  SETTING UP THE MODEL
_____________________________

HOW MANY PRINCIPAL ATTRIBUTES
 5

GIVE PRINCIPAL ATTRIBUTES ALPHABETICALLY

APPEARANCE
COST
PERFORMANCE
RELIABILITY
SAFETY

GIVE PRIMARY DECISIONS

 55
 70
 80
 65

MODEL SOLUTION
______________

 .33 APPEARANCE
 .27 COST
 .18 SAFETY
 .14 PERFORMANCE
 .08 RELIABILITY
```

Figure 1.11 CADEMAK printout of attribute model

lines 450–590 bubble sort, largest values top, lowest values bottom
lines 600–630 printout attribute model, priority listing and ratings.

Before leaving the subject of attribute modelling, it is important to stress that **the use of the five principal attributes is only for the purpose of demonstrating the procedure.**

In real life applications, other attributes additional to the five must be taken into account. For example, applying the technique to the attribute model of a proposed machine tool design might produce this result.

Attribute		Factors to be considered
1	appearance	shape, size, colour, texture
2	capacity	size, force, movement, direction
3	cost	initial, running, maintenance, trade-in
4	interchangeability	rapidity, accuracy, multi-role, modules
5	life	first line, backup
6	maintainability	continuous, regular, sporadic, none
7	performance	force, velocity, pressure, energy
8	reliability	MTTF, MTBF, repeatability
9	safety	electrical, mechanical, chemical, noise
10	simplicity	technology, manufacture, maintenance, use

After prolonged and careful consideration of the above, and also discussion with current machine tool users and potential customers, the following allocation of marks might be agreed.

Attribute	Comparisons									
	1/2	1/3	1/4	1/5	1/6	1/7	1/8	1/9	1/10	Factor
1 appearance	35	14	30	38	30	10	12	15	25	1.000
2 capacity	65									1.857
3 cost		86								6.143
4 interchangeability			70							2.333
5 life				62						1.632
6 maintainability					70					2.333
7 performance						90				9.000
8 reliability							88			7.333
9 safety								85		5.666
10 simplicity									75	3.000

The results of running these values through the computer are shown in Figure 1.12. Figure 1.13 shows a printout of a more crude attribute model using only the five principal attributes, but using the same primary decision values as in the more complete model. In both cases the priority order of the principal attributes is the same, as would be expected. Only the ratings are different, reflecting the use of different numbers of attributes.

```
COMPUTER AIDED DECISION MAKING

PHASE 1  SETTING UP THE MODEL

HOW MANY PRINCIPAL ATTRIBUTES
 10

GIVE PRINCIPAL ATTRIBUTES ALPHABETICALLY

APPEARANCE
CAPACITY
COST
INTERCHANGEABILITY
LIFE
MAINTAINABILITY
PERFORMANCE
RELIABILITY
SAFETY
SIMPLICITY

GIVE PRIMARY DECISIONS

 35
 14
 30
 38
 30
 10
 12
 15
 25

MODEL SOLUTION

 .22 PERFORMANCE
 .18 RELIABILITY
 .15 COST
 .14 SAFETY
 .07 SIMPLICITY
 .06 INTERCHANGEABILITY
 .06 MAINTAINABILITY
 .05 CAPACITY
 .04 LIFE
 .02 APPEARANCE
```

Figure 1.12 Attribute model of machine tool (10 attributes)

```
COMPUTER AIDED DECISION MAKING

PHASE 1  SETTING UP THE MODEL

HOW MANY PRINCIPAL ATTRIBUTES
 5

GIVE PRINCIPAL ATTRIBUTES ALPHABETICALLY

APPEARANCE
COST
PERFORMANCE
RELIABILITY
SAFETY

GIVE PRIMARY DECISIONS

 14
 10
 12
 15

MODEL SOLUTION

 .31 PERFORMANCE
 .25 RELIABILITY
 .21 COST
 .19 SAFETY
 .03 APPEARANCE
```

Figure 1.13 Attribute model of machine tool (5 attributes)

2 Decision Making Explored

Our greatest individual asset is the ability to think creatively. It is our most valuable personal resource and it should be assiduously cultivated.

In striving to improve our creative thought capability, it is important that we avoid the development of habit and prejudice. Such luxuries lull the brain into *automatic pilot* activity and progressively inhibit creativity, through lack of practice. There is nothing wrong with pigeon-holing a good, proven idea, but don't use it so often that your work becomes as recognisable as your signature.

In creating a system to satisfy a need, the designer utilises two basic skills; communication and decision making. Communication characterises those activities concerned with the receiving and transmitting of information. The problem finding function of design is almost entirely communication based.

Decision making is almost wholly confined to the problem solving function, wherein the designer is concerned with making choices between available alternative courses of action. **Decision making is the real workload of the designer, and it is the skill which is least understood.**

The Pareto distribution has already been introduced and, significantly from the point of view of the designer, the concept can be applied to decision making.

In any design, a very few decisions, perhaps less than ten, exert a major influence on the whole design concept. They are basic, their effects are far reaching and supremely important. They set the pattern for the design. They pre-ordain the constraints to be placed upon future decisions, as yet unmade. To a very great extent, these few decisions determine whether the ultimate design will be 'good' or 'bad'.

Having made these very few decisions, the designer will then go on to make a large number of relatively less important decisions, each of which will spring from one of the few, and will be dependent upon it. By this is meant that if the basic decision is changed, the likelihood is that the dependent decision will also be changed.

After these secondary decisions will come a vast mass of almost unimportant and arbitrary choices, each springing from and being dependent upon decisions previously made.

In this way the whole design evolves. The Pareto relationship is apparent: a very few decisions exerting a major effect; many having a middling effect; the vast majority having hardly any effect at all.

These three levels of decision may be categorised as **fundamental, intermediate** and **minor**. They may be likened to the growth of a tree; the fundamental is the trunk upon which all other growth is based, the intermediates are the main boughs and these support the smaller branches and twigs which equate to the minors. Let us look at them in more detail.

Fundamental decisions

It is impossible to stress too highly the importance of these. Not only are their effects of the utmost significance to the whole design concept, but ... **they are always made right at the front end of the project, when the designer's knowledge of the future evolution of the design is a minimum!**

Fundamental decisions must always be approached with caution, and the temptation to 'jump to conclusions' must be firmly resisted. Fundamental decisions can be identified as those which are irretrievable without catastrophic redesign. To abandon one fundamental for another, is virtually to tear up all design work already accomplished and to start again from scratch. The following are typical fundamentals.

car designed for front-wheel as opposed to rear-wheel drive
continuous, rather than intermittent processing
away-from-site prefabrication instead of on-site construction
use of human rather than automatic process control
design for mass- as opposed to batch-production
design based on a particular manufacturing technique or material
design for regional in preference to centralised manufacture

It is important to remember that a handful of fundamental decisions can exert a 70–80% effect on the overall design. An awareness that they exist and can be identified is halfway toward better decision making.

Intermediate decisions

These follow from fundamentals, and are extensions of and supplementary to them. In the event of a change of mind, they can be retrieved with some difficulty, but always very expensively, usually resulting in significant redesign in their own and associated areas. Many intermediates may spring from one fundamental. Or they may spring from other intermediates. And they are not necessarily all of equal importance.

Consider, for instance, a fundamental decision to design a rotary assembly machine on the principle of continuous rotation rather than intermittent indexing.

Arising directly from the fundamental, we have to make an intermediate decision as to how the various product components of the assembly will be fed to the machine heads. Several choices are available.

1 they could be hand-fed by an operator
2 they could be drop-fed from chutes as the heads pass
3 they could be precision transferred by a moving feeder
4 they could be fed by any combination of the above

Once having made this intermediate decision, which determines the method of feeding, we have to decide how the components will arrive at the various feeding points around the machine.

Several alternatives are available. For example

1 if operator-fed, they may come in open tote pans
2 if drop-fed, they may come in vibratory hoppers or magazines
3 if precision transferred, they may come in magazines or in pre-loaded purpose-built feeder mechanisms
4 if fed by mixed methods, they may come in any number of appropriate ways.

In this case, the decision as to how each component comes to its feeder position is an intermediate, but it springs from the intermediate choice of feeder technique, rather than from the fundamental of continuous rotation.

Minor decisions

These occur in their thousands, following the intermediates. Many spring directly from intermediates, or even occasionally from fundamentals, while some come from other minors.

They are most often concerned with design details. They cover such things as component geometry, materials, processes, finishes, tolerances, treatments, etc. For instance, the choice of carbon steel as the material of a component could be a minor decision. Springing from that might come another minor decision as to whether to electroplate or paint the finished component to inhibit rust.

It is in the field of minor decisions that the value analyst employs cost reduction techniques (references 8, 11). By combining components, using less expensive materials and finishes, widening tolerances, eliminating unnecessary features, etc., etc., the cost of the produce can be pruned without adversely affecting its quality. The value analyst rarely operates in the field of intermediates, and never in the area of fundamentals.

Forcing decisions

Thus far, we have developed an ability to set up a model of the projected design in terms of its attributes, and we have looked at the sorts of decisions we may have to make when choosing from a number of alternative possiblities.

But which of the alternatives is the best? How will the designer recognise the best when he or she sees it? Indeed, how can we decide whether any one alternative is better or worse than any other?

Forced decision analysis

Clearly each alternative must be put up against the attribute model and compared with it in order to assess its relative position in the 'league table'. And this comparison need not be an intuitive one. We have the means to make it numerical if we use the pairs-comparison technique.

To see how this might be done, consider an aspect of the vacuum cleaner design. A decision which must be made concerns the storage of dust in the finished design. The dust could be stored externally, or it could be concealed within the framework of the machine.

This will be an intermediate decision. If we subsequently have a change of mind, then it will mean a complete redesign of that part of the cleaner and also of associated parts. Let us opt for external storage. We now face another intermediate decision on the method of storage. Consider a few of the alternatives.

plastic container
fabric bag
paper sack

Each of these alternatives must be compared with the attribute model so that we can assess which is most suitable. Firstly, look at the attribute of appearance.

Attribute + Rating	Alternative	Primary decisions		Comments
appearance	plastic	60	75	good colour and finish range
0.33 (from model)	fabric	40		more limited finish range
good = high mark	paper		25	very limited finish range

Now look at the next attribute, then the next, and so on ...

Attribute + Rating	Alternative	Primary decisions		Comments
cost	plastic	45	20	high cost material and tooling
0.27	fabric	55		medium cost material, low cost tooling
low = high mark	paper		80	low cost material and tooling
satety	plastic	55	85	rugged but will chip
0.18	fabric	45		flexible but will abraid
good = high mark	paper		15	flimsy and will tear
performance	plastic	30	40	needs a separate dust filter
0.14	fabric	70		self filtering
good = high mark	paper		60	self filtering
reliability	plastic	55	75	robust, long lasting, durable
0.08	fabric	45		subject to long term wear and spillage
good = high mark	paper		25	easily torn and contents spilled

From this analysis, provided we are satisfied with the values of the primary decisions, we can proceed to the arithmetic. If we are not satisfied with the primary decisions, we should go back and re-think the values. It is not too late to make changes. Changes made now could avoid problems later on. Let us say we are satisfied with the present values, and press on to the next stage. (See Table on facing page). Let us draw breath and examine this table in detail to see just what we have done.

Consider firstly the attribute of appearance. The first three columns are a repeat of the details on this page. In the fourth column are the factors derived from the primary decisions, as originally explained on pages 21 and 22. Plastic compared with itself is $60/60 = 1.000$, fabric compared with plastic is $40/60 = 0.667$, paper compared with plastic is $25/75 = 0.333$.

Attribute + Rating		Alternative	Decisions		Factor	Normal	Normalised factor × Rating		
							Plastic	Fabric	Paper
appearance	0.33	plastic	60	75	1.000	0.500	0.165		
		fabric	40		0.667	0.333		0.110	
		paper		25	0.333	0.167			0.055
cost	0.27	plastic	45	20	1.000	0.160	0.043		
		fabric	55		1.222	0.196		0.053	
		paper		80	4.000	0.643			0.173
safety	0.18	plastic	55	85	1.000	0.501	0.090		
		fabric	45		0.818	0.410		0.074	
		paper		15	0.176	0.088			0.016
performance	0.14	plastic	30	40	1.000	0.207	0.029		
		fabric	70		2.333	0.483		0.067	
		paper		60	1.500	0.310			0.043
reliability	0.08	plastic	55	75	1.000	0.466	0.037		
		fabric	45		0.818	0.380		0.030	
		paper		25	0.333	0.155			0.012
						Totals	0.364	0.334	0.299

The fifth column shows these factors normalised to unity total. The total of the fourth column is 1.000 + 0.667 + 0.333 = 2.000. And the values for the fifth column are 1.000/2.000 = 0.500, 0.667/2.000 = 0.333, 0.333/2.000 = 0.167. What the fifth column is telling us is that from the total of unity, the primary decisions give 0.500 to plastic, 0.333 to fabric and 0.167 to paper, **for appearance only.**

However, the total rating for the attribute of appearance is only 0.33 as determined by the attribute model, and not unity. So by multiplying each normalised factor by the attribute rating (0.33), we get its own portion of the total attribute rating. These values are shown in the three columns on the right. Overall, for appearance only, the rating of 0.33 has been apportioned by the primary decisions thus—0.165 to plastic, 0.110 to fabric and 0.055 to paper.

By repeating this procedure for each of the remaining attributes, we build up a table of merit for each of the alternatives considered, and these values are shown in the totals at the base of the right hand columns. Our primary decisions indicate that overall the merit rating is...

0.364 plastic
0.334 fabric
0.299 paper

0.997 total

and so we should opt for the plastic container as our form of external dust storage.

```
100 PRINT CLS
110 PRINT"COMPUTER AIDED DECISION MAKING             "
120 PRINT"______________________________  ___________"
130 PRINT:PRINT
140 PRINT"PHASE 2  FORCED DECISION ANALYSIS"
150 PRINT
160 INPUT"HOW MANY ATTRIBUTES ";N
170 PRINT
180 INPUT"HOW MANY ALTERNATIVES ";P
190 DIMA$(N),L$(P),F(P-1),D(P),B(P),Z(P),K(N),Y(N),A(N,P),T(N)
200 PRINT
210 PRINT"ATTRIBUTES"
220 FORI=1TON:INPUTA$(I):NEXTI
230 PRINT
240 PRINT"ALTERNATIVES"
250 FORJ=1TOP:INPUTL$(J):NEXTJ
260 PRINT:PRINT
270 REM CALC OF NORMALISED FACTORS
280 PRINT:PRINT
290 FORI=1TON
300 PRINT
310 PRINT
320 PRINT"RATING FOR ";A$(I):INPUTK(I)
330 IF K(I)>1THEN320
340 PRINT
350 PRINT"PRIMARY DECISIONS FOR ";A$(I)
360 F=1:X=1
370 FORJ=1TOP-1:INPUTD(J):NEXTJ
380 FORJ=1TOP-1:B(J)=100-D(J)
390 F(J)=INT((B(J)/D(J))*1000+.5)/1000
400 X=X+F(J):NEXTJ
410 REM ADD & CONVERT TO UNITY TOTAL
420 Y(I)=INT(((F/X)*K(I))*1000+.5)/1000
430 LETA(I,1)=Y(I)
440 FORJ=1TOP-1:Z(J)=INT(((F(J)/X)*K(I))*1000+.5)/1000
450 LETA(I,J+1)=Z(J)
460 NEXTJ
470 NEXTI
480 PRINT:PRINT
490 PRINT"MERIT RATING OF ALTERNATIVES"
500 PRINT
510 FORJ=1TOP:FORI=1TON
520 T(J)=T(J)+A(I,J):NEXTI
530 NEXTJ
540 REM BUBBLE SORT
550 FORI=1TOP-1:FORJ=1TOP-1
560 IFT(J)>=T(J+1)THEN630
570 LETG=T(J)
580 LETV$=L$(J)
590 LETT(J)=T(J+1)
600 LETL$(J)=L$(J+1)
610 LETT(J+1)=G
620 LETL$(J+1)=V$
630 NEXTJ,I
640 PRINT:PRINT
650 FORJ=1TOP:PRINTT(J),L$(J):NEXTJ
READY.
```

Figure 2.1 Forced decision analysis program

```
COMPUTER AIDED DECISION MAKING

PHASE 2  FORCED DECISION ANALYSIS

HOW MANY ATTRIBUTES  5

HOW MANY ALTERNATIVES  3

ATTRIBUTES
APPEARANCE
COST
SAFETY
PERFORMANCE
RELIABILITY

ALTERNATIVES
PLASTIC
FABRIC
PAPER


RATING FOR APPEARANCE
 .33
PRIMARY DECISIONS FOR APPEARANCE
 60
 70


RATING FOR COST
 .27
PRIMARY DECISIONS FOR COST
 45
 20


RATING FOR SAFETY
 .18
PRIMARY DECISIONS FOR SAFETY
 55
 85


RATING FOR PERFORMANCE
 .14
PRIMARY DECISIONS FOR PERFORMANCE
 30
 40


RATING FOR RELIABILITY
 .08
PRIMARY DECISIONS FOR RELIABILITY
 55
 75


MERIT RATING OF ALTERNATIVES

 .356           PLASTIC
 .33           FABRIC
 .313           PAPER
```

Figure 2.2 Printout for solution of dust storage problem

If this choice does not agree with that which we would have made intuitively, it does not necessarily mean that there is something wrong with this method. What it does mean is that there is a difference between our intuitive reasoning and the more detailed analytical decision making of this method. What we have to do is decide which process is at fault. In all probability it is the intuitive reasoning which is suspect, because it can only make decisions at a fairly superficial level, while pairs-comparison forces detailed examination of all the aspects before a decision is reached.

For a fairly limited range of alternative choices, we have been involved in quite a lot of calculation, and this could escalate for a wider range of alternatives. Here, once again, we can turn to the microcomputer and offload the humdrum work. As before, the main parts of the program are

lines 100–160 title of the program
lines 170–380 data input, attributes, alternatives, ratings
lines 390–490 compute factors, add, convert to unity total
lines 500–640 bubble sort, high values top, low values bottom
lines 650–660 display merit ratings and alternatives.

Figure 2.1 shows the complete program and Figure 2.2 is a printout for the dust storage problem.

This decision making analysis is based on the best evidence we have at the present moment. It takes our subjective assessment of the ideal attribute model and combines it with our, also, subjective assessment of the various values derived on pages 32 and 33. So the whole exercise is subjective. But so is every decision of designers. They take the best evidence available, analyse it, and come to a conclusion with which they must live, and which must be defended against the logic of others. It is tough, but it is the lot of the designer.

Summary

Thus far we have developed a couple of computer programs which enable us to suggest a priority listing and award ratings to a number of design attributes, and then to select from a range of alternatives that one which most closely matches those attributes.

However, there is another aspect of the forced decision analysis which should be mentioned. Our original attribute model for external dust storage on the vacuum cleaner had a profile like this ...

0.33	appearance
0.27	cost
0.18	safety
0.14	performance
0.08	reliability
1.00	total

The alternative of plastic, which we selected by the forced decision analysis, has a profile like this ...

0.165	appearance—which normalises to unity total as	0.45
0.043	cost	0.12
0.090	safety	0.25
0.029	performance	0.08
0.037	reliability	0.10
0.364		1.00 total

and this is not a very close profile match with the original model. Clearly, appearance is heavily overweighted, while cost and performance are each only rated at about half of what the original model requires.

If we examine the other two alternatives of fabric and paper, we see that they have the following profiles ...

fabric			paper		
0.110	which normalises to	0.33	0.055	which normalises to	0.16
0.053		0.18	0.173		0.58
0.074		0.22	0.016		0.05
0.067		0.20	0.043		0.14
0.030		0.09	0.012		0.04
0.334		1.00	0.299		0.99

While all three have variances from the original model profile, it is clear that the alternative which most nearly matches is that of fabric, and perhaps this would be a better choice. However, although theoretically correct, this approach is perhaps 'gilding the lily' unnecessarily. The whole of the pairs-comparison technique is very subjective, and elaborating to this level of detail is a clear case of diminishing returns. For forcing decisions between alternatives, the overall merit rating is good enough as an initial guide, and common sense must always be used in its application.

This type of analysis may be used whenever a choice is obscure and the final decision is not clear. It should certainly be used when making any fundamental and some intermediate decisions. But not for minor decisions; intuition is usually accurate enough for these, and very much quicker.

3 Manufacturing Resources

From the previous comments on decision making, it is clear that making correct decisions requires more than just good intentions. It requires a knowledge of materials, manufacturing processes, labour skills, technology levels, costs, quality, and many other things. The broader this knowledge, the more likely that good decisions will be made.

In making decisions, the designer is determining which manufacturing resources will be used to create the design, and also how those resources will be manipulated. Only five resources are available, those which were considered when setting the parameters of the design solution, viz ...

labour	skill requirements, technology levels
materials	manufacturing materials, energy, equipment
time	timescale for implementation
space	where implementation will occur
money	cost of implementation

Each of these resources can be broken down further as follows ...

labour	direct (productive), indirect (supportive)
materials	revenue (consumable), capital (durable)
time	preparation (get ready), manufacturing (produce)
space	active (working), passive (storage)
money	revenue (consumables, wages, etc.), capital (buildings, services, plant, etc.)

Direct labour is that of people whose efforts can be recognised in the end product. Typically, in the vacuum cleaner this would include foundrymen, plastic moulders, machinists, coil winders, assemblers, inspectors, testers, etc.

Indirect labour is that of people who are used in general support of manufacturing, but whose efforts are not recognisable in the finished product. Typically, this would include labourers, storekeepers, supervisors, managers, office personnel, designers, planners, purchasers, accountants, canteen staff, transport drivers, medical staff, etc.

Revenue materials are those which are consumed in the making of the product, and would include plastics, metals, wire, bearings, screws, etc., and which, like direct labour, can be identified in the finished product.

Capital materials are those which are required to support manufacture but which cannot be identified in the finished product. They would include buildings, plant, tooling, services such as power, phone, telex, computers, etc.

Preparation time is all that spent in getting ready for manufacture, such as market surveys, development, design, tool manufacture, planning, materials procurement, factory layout activities, equipment commissioning, etc.

Manufacturing time is self explanatory and includes all production activities, plus inspection, testing, packaging, etc.

Active space is all that used in manufacturing operations and may include foundries, moulding shops, machine shops, assembly areas, inspection and test bays, etc.

Passive space is all the rest and includes some areas closely allied to manufacture, such as main and auxiliary storage, work-in-progress areas, kit marshalling compounds, packing bays, etc.

Money is the resource which is common to all the above items. Revenue expense covers all expenditure related to manufacture and will include consumable materials, wages, salaries, energy costs, recycled material costs, receipts from end-of-life trade-ins, etc., etc.

Capital expense covers buildings, services, plant, insurance, etc.

The decisions which the designer makes will directly affect which resources will be needed, where they will be used, and for how long. For instance, if the designer creates a simple cylindrical sleeve for use as a spacer on the vacuum cleaner; immediately the shape, the material, and the nominal size of this component is decided, this dictates some of the resources which will be used in its manufacture, such as

processes	technology, labour class
plant	machine capacity, working areas
skill levels	labour grades

When the designer assigns tolerances, treatments, manufacturing quantities, further resources will be committed, such as

class of machinery, accuracy, repeatability
sophistication and types of tooling and inspection aids
detailed manufacturing processes, materials movement
manufacturing methods, operational planning
active and passive space, manufacturing, storage, work-in-progress, kit marshalling
make-or-buy decisions, in-house manufacture, subcontractors
proprietary items for procurement, component stocking policy
recruitment and training needs
services requirements, power, maintenance, canteen, etc.
and many more.

From this it can be seen that the decisions of the designer have very far-reaching effects. As the purpose of any engineering undertaking is to make a profit in order to be able to stay in business, the decisions of the designer will be crucial to the well-being of the company. So the designer had better be sure that decisions are cost-beneficial.

Of the total of five resources, two have to be manipulated with extreme care. These are the resources of time and money. If you run out of either before the project is complete, then the project fails, and maybe so does the company. So we will turn our attention to the proper husbandry of time and money budgets.

The resource of time

In all human activities, time plays an important role. All activities in which we take part have a start, a middle, and a finish. In order to make the best use of time, it is essential to plan and monitor its usage. And this applies especially to design activities.

There are several accepted ways of planning and monitoring the use of time, all of which employ the following...

set out detailed plan of all activities
decide checkpoints (milestones where actual progress can be compared with the master plan)
exercise control by periodic review of actual progress
adjust checkpoints and targets as necessary.

This usually results in a graphical display of the overall plan, known as a **network,** which shows the principal activities and their inter-relationships, one with another. A typical network could be a bar chart (see Figure 3.1).

Or the network could take the form of a critical path analysis (Figure 3.2).

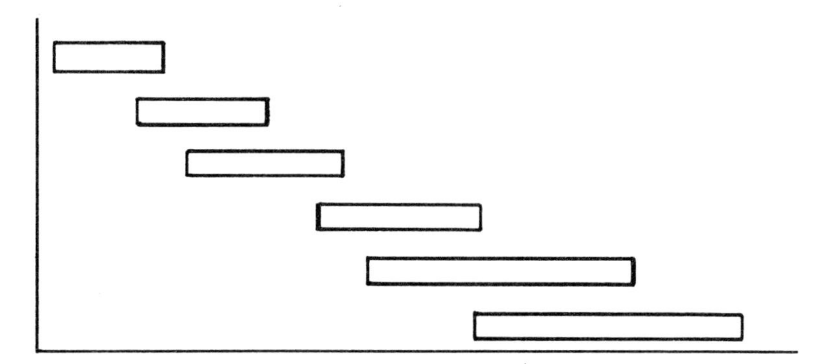

Figure 3.1 Typical bar chart

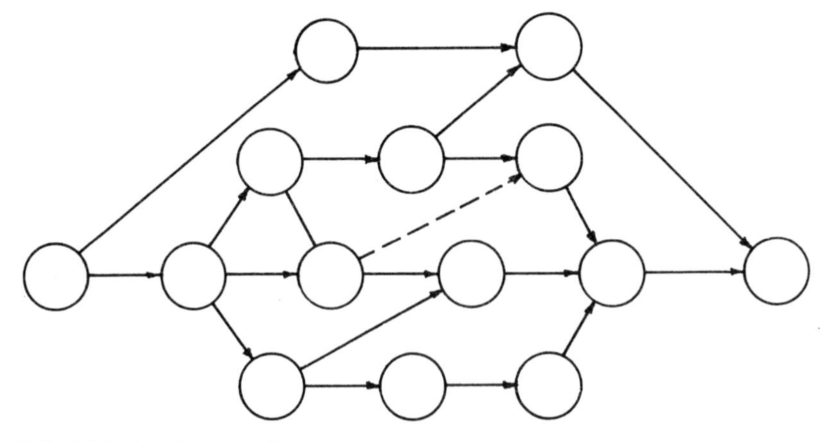

Figure 3.2 Critical path network

Critical path analysis

Of the two networks, the critical path type is far more powerful than the bar chart, because it shows detailed inter-relationships between activities in a more meaningful way. It is probably the most powerful management planning technique to have been developed in the past fifty years. To illustrate its potential, consider a very simple, non-industrial example.

Three men leave their place of work at 12.30 p.m. after Saturday morning overtime, having decided to go to a football match two miles from the works. Kick-off is at 12.45 p.m. Each man has a different route, or mode of travel, and they agree to meet outside the ground and all go in together.

Thus, the project begins with their finishing work at 12.30 p.m. and is to be completed with their meeting outside the ground in time for kick-off, fifteen minutes later.

The planning phase of the project reveals the following details.

Man 1 will cycle directly to the ground in 12 minutes.

Man 2 will walk to a tobacconist kiosk for cigarettes in 5 minutes, and then walk to the ground in a further 25 minutes

Man 3 will also walk to the same kiosk in 5 minutes, then to his home in a further 15 minutes to let his wife know his plans, thereafter he will walk to the ground in a further 30 minutes.

The scheduling phase of the project shows the sequence of activities and their inter-relationships in network form.

The network is made up of activities and events. An activity is represented by an arrow indicating the consumption of a resource, in this case time. The head of the arrow points in the direction of the resource consumption.

An event, which is merely an identifiable point in time, is represented by a circle, divided as shown in Figure 3.3 for the purpose of providing relevant information about the event. The upper half contains the event number, while the lower quadrants contain, on the left the earliest time for event occurrence and on the right, the latest time for event occurrence. In this particular network, all times are in minutes.

Earliest event time is calculated by adding activity duration time to the earliest event time applying at the preceding event. Thus, starting from the beginning of the network, event 1 has an earliest event time of zero. The earliest time for event 2 is zero plus 5 minutes, the duration of the activity joining events 1 and 2. This activity is designated 1–2.

Earliest time for event 3 is 5 plus 15 minutes for activity 2–3.

Earliest time for event 4 is 20 plus 30 minutes for activity 3–4, or so it would appear. However, if we look at the other paths through the network we see that the earliest time for event 4 can also be 5 plus 25 minutes for activity 2–4. Similarly, it can also be zero plus 12 minutes for activity 1–4. There appears to be a contradictory state of affairs.

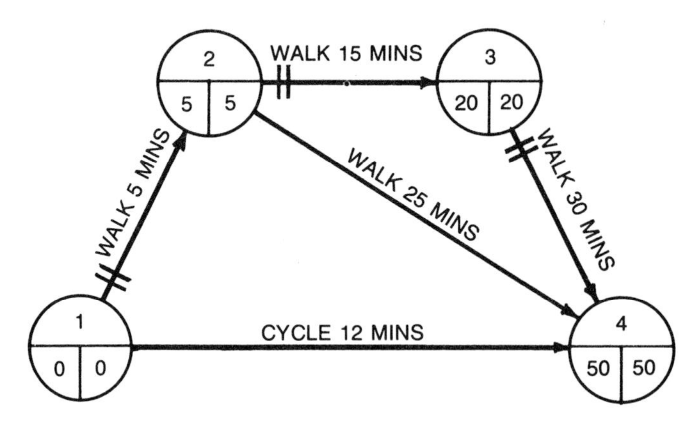

Figure 3.3 Critical path network — initial

It must be remembered that we are showing the inter-relationships that exist between disparate activities which together make up the complete network. Clearly, the shortest time in which this network of activities and events can be completed, is determined by the *longest* time-path through the network, from event 1 to event 4. And this, the **critical path**, is given by the activities 1–2 + 2–3 + 3–4, 50 minutes in all. And this is the shortest time in which the network can be completed, unless we can reduce the time of activities on the critical path, or rearrange the network order to give a shorter critical path.

Once the earliest time for event 4 has been calculated, the latest time for that event will also be the same value, i.e. 50 minutes. The latest times for all other events can now be calculated by working backwards through the network from event 4 to event 1, looking always for the longest path to every event.

Let us set out the network in tabular form.

Table 3.1 shows the details of each activity in terms of its identity, duration, earliest and latest start and finish times, total float, and activity description.

Total float is latest start minus earliest start, and shows the amount of available slack in the activity. Thus, activity 1–4 could start at any time between zero and 38 minutes and still finish within the network completion time of 50 minutes, without upsetting any other activity. This float can be used to our advantage.

Table 3.1 Critical path schedule — initial

		Start		Finish			
Activity	Duration	Earliest	Latest	Earliest	Latest	Total float	Activity description
1–2	5	0	0	5	5	0	walk from work to kiosk
2–3	15	5	5	20	20	0	walk from kiosk to home
2–4	25	5	25	30	50	20	walk from kiosk to ground
3–4	30	20	20	50	50	0	walk from home to ground
1–4	12	0	38	12	50	38	cycle from work to ground

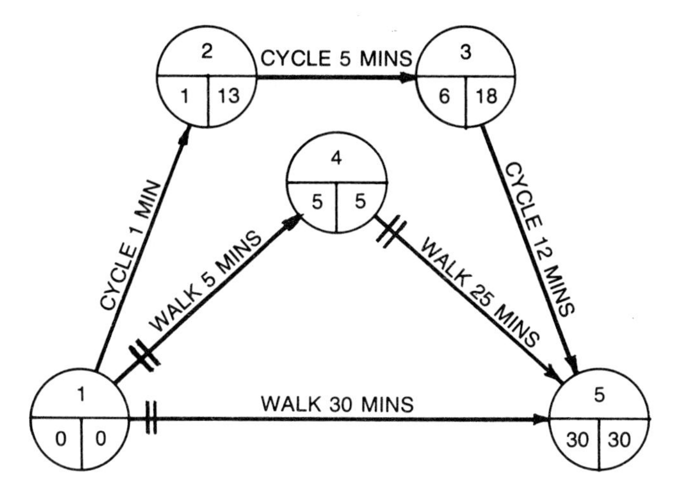

Figure 3.4 Critical path network — intermediate

Activities with zero total float form the critical path through the network, i.e. 1–2, 2–3, 3–4. Any slip in completing critical activities, means an equivalent slip in network completion time. In the network, critical activities are identified by two short cross bars.

Let us return to the three men and their football match. Arrival of the last of the three at the ground some 35 minutes after kick-off means they will miss most of the first half of the game. Remember, they had agreed to meet at the gates and all go in together. Clearly, we have to find a way to reduce the overall duration of the critical path. This means reducing some or all of the critical activities, or rearranging the network for earlier completion. Firstly, consider an internal transfer of resources from non-critical to critical activities (Figure 3.4).

> Man 1 offers his cycle to man 3. Man 1 will now walk from the works to the ground in 30 minutes.
>
> Man 3 will now cycle from the works to the kiosk in 1 minute, then to his home in 5 minutes, and finally cycle to the ground in a further 12 minutes.
>
> Man 2 is unaffected by these changes and will still walk from the works to the ground, via the kiosk, in a total time of 30 minutes.

Whereas before, man 2 and man 3 arrived at the tobacconist kiosk simultaneously, they now get there 4 minutes apart, so the event occurs twice and has to be recorded twice, as events 2 and 4.

It will be seen that this simple internal transfer of resources has reduced the overall time for network completion from 50 minutes to 30 minutes. It will also be noticed that there are now two critical paths — activity 1–5 and activities 1–4 + 4–5. However, we have made a very significant reduction in the overall network time, but not enough for the men to be present at the start of the match.

It is now necessary to increase the resources available to the three men if we are to reduce the overall project time enough for them to get to the ground before kick-off.

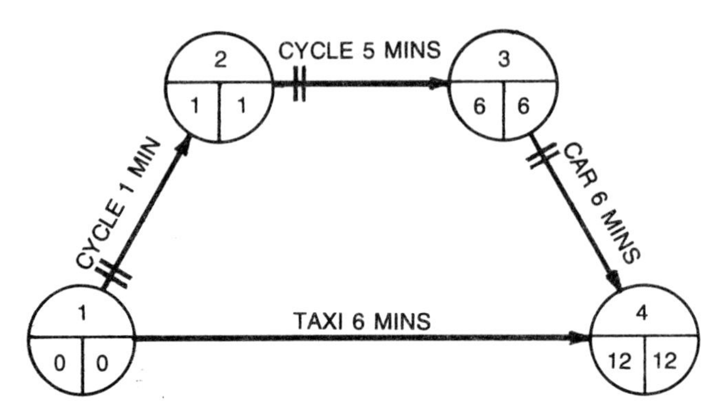

Figure 3.5 Critical path network — final

Man 3 has a car at home. As he cycles from the works, he can stop at the tobacconist kiosk and buy cigarettes for himself and for man 2. His total cycle journey home still takes 6 minutes as before but, once home, he can transfer to his car and reach the ground in a further 6 minutes, making a total trip time of 12 minutes.

Meanwhile, man 1 and man 2 hire a taxi and are driven directly from the works to the ground in 6 minutes. Now the network looks as shown in Figure 3.5, with man 3 back on the critical path.

The important point to notice is that the utilisation of extra resources, together with the internal resource transfer, has enabled the completion of the project within the prescribed timetable. All three men arrive at the ground by 12.42 p.m. three minutes before kick-off and with enough time to get their tickets and to reach their places to see the full match. So by the expedient of rearranging internal resources and also bringing in external resources, the project timescale has been dramatically reduced from 50 minutes to 12 minutes. But at what cost?

Network costs

The cost of reducing project time from 50 minutes to 12 minutes can be established by comparing the total project costs incurred in both the initial and final networks. To do this we must apportion realistic values to all the resources used, both human and non-human. For comparison purposes, we can make the following assumptions...

each man's time is worth 5p per minute
wear and tear on the cycle is 3p per minute
cost of running the car is 10p per minute
cost of hiring the taxi is 30p per minute

Applying these values to the initial network gives

1–2 = 2 men × 5 mins × 5p	50p
2–3 = 1 man × 15 mins × 5p	75p
2–4 = 1 man × 25 mins × 5p	125p
3–4 = 1 man × 30 mins × 5p	150p
1–4 = 1 man + cycle × 12 mins	96p
total cost of project	£4.96

The cost of the final network is

1–2 = 1 man + cycle for 1 min	8p
2–3 = 1 man + cycle for 5 mins	40p
3–4 = 1 man + car for 6 mins	90p
1–4 = 2 men + taxi for 6 mins	240p
total cost of project	£3.78

So it is actually £1.18 cheaper to employ expensive resources in order to reduce the project time by 76%. Although this is a contrived example, it does highlight the important features of critical path techniques is a very simple straightforward way. It also suggests that by reducing the expensive labour content in a project, the use of very expensive equipment can be justified, because of the shorter time for which it is required.

Three-time estimating

But how, you may ask, does all this affect the designer? Most designers would claim that it is not possible to give accurate estimates for what is primarily creative work. After all, the really good idea may come in a very short time, or it may take several weeks to happen.

To overcome this problem we use a technique known as three-time-estimating. We ask the designer three questions in the following sequence.

Firstly, we ask for an estimate of the shortest timescale for project completion, assuming that all decisions yet to be made are correct first time (the optimistic time).

Secondly, we ask for an estimate of the longest timescale for project completion, assuming that every unmade decision will be wrong first time and that perhaps 50% of them will be wrong the second time (the pessimistic time).

Thirdly, we ask for a review of the designer's knowledge of the present state of the project, with all the possible problem areas, and then to give a best assessment of the most likely time for project completion (the most likely time).

Only one time can be used for the duration of the design activity, and this is compiled from the three-time-estimates as follows ...

$$\text{expected time} = (\text{optimistic} + 4 \times \text{most likely} + \text{pessimistic})/6$$

This technique of three-time-estimating has been shown on many projects to be remarkably accurate, as the following true example will illustrate.

A large electrical company had contracted with a Government ministry to design and manufacture test equipment for use with radar cathode ray tubes. The crt's would be stored for up to two years before issue to the services and it was important to carry out exhaustive tests just before issue to ensure no storage deterioration had occurred. Because of the conditions obtaining in the underground storage depot, and the high voltage, >20 kV, on the tubes during test, extreme safety features had to be built into the test equipment to ensure operator protection. For this reason the design moved into technological regions not previously explored. The contract was to be completed in one year, and it contained a punitive penalty clause to ensure on-time delivery.

After seven months work, the development of the equipment was still incomplete,

PROJECT: FOOTBAL MATCH – INITIAL NETWORK

EVENT TIMES

NUMBER	EARLIEST	LATEST	FLOAT	
1	0	0	0	CRITICAL
2	5	5	0	CRITICAL
3	20	20	0	CRITICAL
4	50	50	0	CRITICAL

COMPLETION TIME = 50 UNITS

PROJECT: FOOTBAL MATCH – INTERMEDIATE NETWORK

EVENT TIMES

NUMBER	EARLIEST	LATEST	FLOAT	
1	0	0	0	CRITICAL
2	1	13	12	
3	6	18	12	
4	5	5	0	CRITICAL
5	30	30	0	CRITICAL

COMPLETION TIME = 30 UNITS

PROJECT: FOOTBAL MATCH – FINAL NETWORK

EVENT TIMES

NUMBER	EARLIEST	LATEST	FLOAT	
1	0	0	0	CRITICAL
2	1	1	0	CRITICAL
3	6	6	0	CRITICAL
4	12	12	0	CRITICAL

COMPLETION TIME = 12 UNITS

Figure 3.6 CPA printouts for football match networks

and with a two months building programme not yet started, anxiety as to the probable profitability of the contract began to rise. Also the pressure to finalise design and development became intense.

The senior design engineer concerned was reluctant to commit himself to a firm date for design completion because of the unknown areas ahead. It was at this point that the three-time-estimate technique was applied, with the full cooperation of the senior design engineer, who gave the following estimates.

optimistic time	6 weeks
pessimistic time	26 weeks
most likely time	10 weeks

These figures resulted in an expected time of 12 weeks, and the design was actually completed three days ahead of this date, enabling the whole contract to be completed without penalty.

There are many items of inexpensive commercial software available which will handle simple critical path analysis, and it is better to make use of these rather than to try to develop a homegrown version. The printouts which follow are typical of such programs. They show the three football match networks with which we have dealt in this chapter.

It will be noticed that this version of critical path analysis is event-orientated. However it contains all the relevant detail.

4 The Resource of Money

Product costs originate in design. Note that it is product costs and not production costs that come directly from the designer. Even the most cost-effective design can subsequently be made unnecessarily expensive by uneconomic manufacturing methods. But the irreducible cost of the product is determined by the designer, **and nobody else**.

A poor design decision may well commit a company to twenty or more years of loss-making production, whereas a better conceived decision could well yield twenty or more years of increased profit. Every design decision carries either a cost benefit or a cost penalty, and it is the responsibility of the designer to exercise the greatest care in decision making. A design which fails to meet its cost specification is no better than one which fails to satisfy its technical specification. Furthermore, when all other features of alternative products are equal, the decision by the customer as to whether or not to buy, is largely determined by expense.

In today's world the accountant is much more likely to rise to a position on the board of directors of a company than is a designer. As a result, the decisions of management may be based more on what is economically viable than on what is technically feasible or desirable. But the designer cannot for long escape from becoming closely involved in cost aspects of the business of his or her company. Money is rapidly ousting mathematics as the common language of engineering and the designer, to be effective, must be able to understand and communicate in the language of cost.

The curves in Figure 4.1 show the relationship between manufacturing costs, selling cost, profitability and technical merit. The total cost of producing and distributing a product is made up of the manufacturing cost plus the selling cost, and this total is not constant. Any shift in technical merit will be reflected in total cost. Technical merit may be broadly equated to product quality (reference 12).

If product quality falls, there will be a rise in the selling cost. The sales force will need to put more effort into their dealings with customers who may be reluctant to buy a product which does not come up to their expectations. More complaints will be raised and more manhours will be needed from the marketing and servicing sections of the company. There will be a general loss of customer goodwill which can only be retrieved by additional expenditure on, for instance, special discounts, free gifts, improved servicing arrangements, reduced prices, etc. All these non-standard expenses will reduce profitability, but they will be necessary if the lower quality product is not to damage future business prospects permanently.

On the other hand, if product quality is allowed to rise unchecked, there will be a rise in total cost due to the sharply increasing cost of manufacture. An ill-considered increase in quality standards, perhaps by inspection personnel, can result in completely acceptable components being scrapped or reworked quite unnecessarily. A dozen such

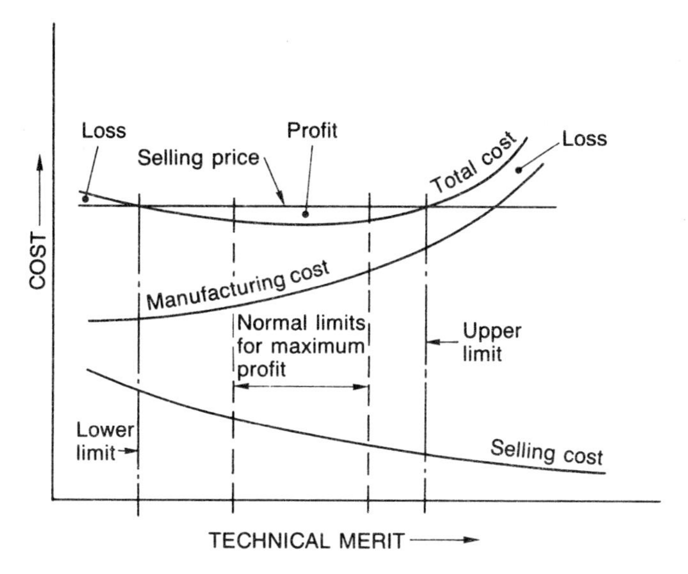

Figure 4.1 Basic cost relationships

marginal shifts of quality in a multi-component product will quickly erode the profit margin and produce a total cost which is greater than the product selling price. From this it will be seen that any excursion beyond the upper or lower limits of technical merit will rapidly turn a condition of profit into one of loss.

The nature of cost

There are several elements which make up the cost of any product. Before examining them in detail, it is important to understand the difference between cost and price.

Cost = expenditure = cash outflow
Price = income = cash inflow

There are three general equations covering these items.

manufacturing cost = direct materials + direct labour + overhead expense (1)

cost of sale = manufacturing cost + amortised special-to-type expense (2)

selling price = cost of sale + sales and administration expense + profit (3)

Note! In equation (2) **cost of sale** is that cost at which the product is transferred from the manufacturing divison to the selling divison of the company.

In equation (3) selling price is the figure at which the product is transferred from the company to its wholesaler, or retailer, or customer, depending upon company distribution policy.

Let us examine the individual items in these equations in more detail.

Materials

The cost of materials will include all items which are charged into the accounts of the company as **material**. There are generally three main groups (reference 3).

1 Raw materials. These are all materials which will be converted into finished or part-finished components on company plant and equipment. Typically, sheet, strip and bar metals, plastics powder, ingots for castings, insulated wire, drawn wire, insulated sheet materials, tubing, timber, fabric, paper, etc., etc.

2 Sub-contract materials. Components and subassemblies made to company part numbers by sub-contract vendors, either because capacity or expertise is not available within the manufacturing facilities of the company, or because the need for these services is sporadic and would not justify the acquisition of the necessary plant or expertise. Items might include castings, forgings, mouldings, electroplating, heat treatment, sintered components, machined components, items bought part-finished to be later completed in-house, special extrusions, etc., etc.

3 Proprietary material. Standard items purchased from specialist vendors, for which company purchasing specifications exist. Typically, electric motors, ball and roller bearings, sleeve bearings, transistors, integrated circuits, printed circuit board blanks, filter units, gear boxes, power supply units, nuts, bolts, washers and screws, springs, pins, dowels, belts, chains, etc., etc.

Although materials are classified as a separate group for costing purposes, their cost content is wholly made up of someone else's labour charges. Consider the processes by which raw material becomes, say, the hub cap of a motor car. Raw material is worked by people, who add the cost of their labour to the cost of the raw material, before it becomes an end product, either for sale to the public or to another manufacturer for further work to be done on it. Table 4.1 shows this process.

The end product of one industry, or of one section of industry, becomes the raw material for another, next in the chain of manufacture.

Since the rock in the ground has no *intrinsic* value, all material costs are the results of labour operations. Material by itself has no value until we set men to dig it up, refine and process it, and transport it to us for conversion into finished products. And this includes energy sources.

Table 4.1 Material becomes an end product

Raw material	+ Human labour	= End product
ore bearing rock	+ mining and crushing	= iron ore
iron ore	+ smelting and refining	= pig iron
pig iron	+ alloying processes	= steel ingots
steel ingots	+ rolling and slitting	= steel sheets
steel sheets	+ pressworking	= components
components	+ treatments	= hub caps

Labour

The labour cost which appears in the manufacturing cost is **direct labour**, i.e. that where the work of the individual man or woman can be recognised in the finished component or product. It will include labour costs for such operations as casting, moulding, machining, fabricating, assembling and testing. It may also include the labour cost of inspection and packaging, but this will depend upon individual company policy. There are probably as many different costing systems as there are separate companies, and each will use a system best suited to its needs.

Cost centre rate

The labour cost for each manufacturing operation is assessed by determining the time to be taken for the operation, and then multiplying this time by an appropriate cost centre rate. Most companies publish their cost centre rates for the use of their technical staff, but a short note is included here as general background information.

Because of the wide differences in the labour categories employed in any one company, it is necessary that they should receive different levels of pay. For example, a highly skilled toolmaker commands a higher rate of pay than a copy typist, and a female personnel officer comes more expensive than a junior storekeeper. The reasons are clear; the law of supply and demand, the type and length of training required, the pressures of trades union negotiations, all make for wide differentials in wage levels.

Direct labour groups may be positioned within a company in a number of ways. They may be grouped together by skills as, for instance, all foundrymen in one area — the foundry. At other times it may be more sensible to form mixed groups of skills based upon the requirements of the manufacturing processes. In such a case, each labour skill will have its own personal level of wage payment, and there will be no common wage level for the group.

Such a grouping of differing skills, each with its own level of wage payment, gives rise to the concept of a cost centre. In a cost centre different wage levels may be paid to individuals within the group, but the hourly charge for the work of the group as a whole is a uniform figure for costing purpose.

For example, consider an assembly group building a small electronic product. Such a group might contain two instrument fitters and three wirewomen working together on the assembly processes. The individual wage levels might look like this.

2 instrument fitters at £2.90 per hour and
3 wirewomen at £1.80 per hour.

Fitters wages are 2 × £2.90 = £5.80 per hour
Wirewomen wages are 3 × £1.80 = £5.40 per hour

Together, this gives a group hourly wage of £11.20. Dividing this group hourly wage by the number of people in the group, gives an average wage of £2.24 per hour for the group as a whole, and this figure will remain unchanged so long as the **mix** of skills within the group does not alter.

To compile the group cost centre rate, we add to this group wage rate the fixed and variable overhead charges which are applicable to the assembly area. The factors which make up these overhead charges will be outlined immediately after the present section.

Typically, a cost centre rate for this assembly group might look like this

average hourly wage rate	£2.24
variable overhead rate	£1.80
fixed overhead rate	£1.55
group cost centre rate per hour	£5.59

This is the amount which will be charged for every hour which the group works. It contains no material costs, only labour and overhead expense. Cost centre rates for all company activities will be similarly structured. If a company does not publish its cost centre rates, the above type of reasoning may be used to establish probable rates, for the comparison of costs of alternative designs. It is not essential that rates used in such comparisons are absolute, but they should enable relative costs to be assessed.

Overheads

These are indirect expenses, both fixed and variable, which are incurred in the manufacture of the product but which cannot be readily identified in the finished article (13).

Fixed overhead is the apportionment of all non-varying expenses, i.e. rent and rates, depreciation on buildings, factory services installations such as power, telephone, telex, etc., general management such as financial, personnel, marketing, technical — including the cost of designing, etc. Fixed overheads may be apportioned on a basis of area occupied, or on numbers of people employed in particular manufacturing activities, or on an arbitrary basis, or on any other basis which is meaningful to the company for the recovery of fixed indirect expense.

Variable overhead is the sum of all indirect expenses which vary approximately in step with the volume of manufacturing work undertaken. It will include supervision, maintenance costs, indirect production workers such as storemen, labourers, service operators, and also the depreciation on manufacturing plant, etc.

Any expense which cannot properly be charged directly to a particular product, may find its way eventually into overhead expense, and will be recovered by being spread over the manufacturing areas of the company. Each company has its own interpretation of what is fixed and what is variable, so the above explanation is general in form only.

Special-to-type expenses

These are all those costs incurred in bringing the manufacturing project to fruition, but which are not accounted for in the manufacturing cost itself. They will usually include at least the following items.

1 Test equipment. All equipment required for the testing and commissioning of the product, making proper allowance for the anticipated maximum throughput quantities, which may require multiple testing facilities. These items will be only

those to be used on the particular product and no other. They will not include any of the items of equipment which the company must have in order to operate its business in a general way. For example, a purpose-built test rig for carrying out final checks on one specific product, would be classified as special-to-type. A motor-generator set designed to deliver special voltage ranges and for use generally in the testing of all products, would be classified as general-purpose equipment and would be paid for from the test department's capital budget.

2 Tooling. Costs of any tools required for the manufacture of components on the company premises, again only of a special-to-type nature. Wherever sub-contractors make tools for their own use in supplying components to the company, and where such tools are invoiced separately from those components, the cost of such tools will be included here. Tooling costs may also include casting patterns, inspection fixtures and gauges, and also assembly and commissioning aids.

3 Technical support. Any project-oriented costs incurred in support of the project, both prior to and during early manufacture, will be included here. Project engineering, coordination, process planning, layout, special engineering work, network analysis for project control, all may qualify for inclusion here.

4 Excess costs. Preproduction and early manufacture will attract material and labour costs in excess of those standard costs estimated in the manufacturing cost of the product. This is due to all manner of problems arising during the familiarisation phase of any new design. Typically, tools may not produce the required components within tolerances, and these items may have to be reworked, thus attracting an extra labour charge. Alternatively, they may have to be scrapped, in which case the whole material and labour content of the rejected item is lost, but this lost cost still has to be paid for.

Special-to-type expenses are normally recovered during the manufacturing life of the product, or over any shorter period of time which management may determine. Clearly, this expenditure is quite high during the preproduction and early manufacturing phases, but will fall significantly as manufacture becomes established and stabilised. The method of recovery is to estimate the probable total special-to-type expense for the new product, then divide this sum of money by the total number of units of the product which will be produced over the recovery period. The resulting figure will represent the additional cost, over and above the standard cost, of producing these units, and this figure will be added to the cost of sale while these units are being manufactured. This procedure is called **amortisation**. Some companies arrange to amortise all special-to-type expenses in the first year or two of new product manufacture, which then allows them to maintain constant selling prices in an inflationary climate or, hopefully, offer their customers a price reduction after completion of amortisation. This may help stimulate demand for the product at a time when the competition is entering the market with a similar product. Such decisions will depend largely on rate of inflation, type of product, rate of obsolescence of the product, degree of competition, and other factors.

Sales and administration expenses

The **mark up** which the sales division adds to the cost of sale is to recover the expense of operating the marketing process. This will include advertising, initial sales, customer liaison, promotion, after-sales service, warranty costs, etc. Normally this mark up is computed as a percentage of the cost of sale of the product, a figure between 5% and 20% being applied, depending upon the total costs of the services involved.

Profit

This will depend entirely upon the type of market within which the company is operating. In highly competitive conditions, a very small margin is all that can be achieved if competitive edge is to be maintained. On the other hand, in a market which is new and expanding and where insufficient of the product can be produced to satisfy customer demand, profit may well be set unnecessarily high to exploit boom conditions while they last. Profit margins may vary from a few percent to hundreds of percent, across this spectrum.

Cost prediction

Whenever a manufacturing company considers the introduction to its range of a new product, it must know as early as possible whether the proposed product will match, or beat, its present competition in both performance and price.

It follows, therefore, that a number of important decisions concerning technology and cost will have to be made before a settled design exists, so that the decision whether or not to proceed with product launch can be based on the best information available. The accurate prediction, in advance of firm design, of probable performance and cost levels will help management to decide manufacturing potential. It will also do something else. If the performance and cost predictions favour product launch, then the designer has ready-made performance and cost targets, from these predictions, within which the design must evolve. However, if the predictions are against product launch, then the designer has the basis, also from the predictions, for performance improvement and cost reduction exercises on the product. Cost prediction is not based upon uninformed guesswork, neither does it come from detailed cost estimates for, as yet, there is no detailed design. It is achieved by three actions.

1 maximising all known factors
2 anticipatory decision making
3 informed guesswork in areas of uncertainty.

Known factors may include knowledge of competitive products, market requirements for quantities, price levels, quality standards, product life requirement, processes to be used in manufacture, etc., etc.

Decisions which have to be anticipated may embrace the value of, and amortisation period for, special-to-type tooling and equipment, amount and disposition of excess material and labour costs, degree of technical support, make or buy policy, etc., etc.

Informed guesswork is kept to a minimum. It will be used to establish the probable cost relationships between component parts of the product. Clearly, if the new product is similar to an existing range of products, then the amount of actual guesswork is very small indeed, and the accuracy of the guesses should be quite high.

The ball-park concept

For an initial cost prediction, preparation time is extremely short. Of necessity, the information upon which such a prediction is based will be incomplete, for it is compiled far in advance of firm design work. Hence, the prediction is likely to be wide of the true figure — as wide as a ball-park.

However, the designer does not start with a completely blank sheet. Whatever the proposed project may be, he or she will have some notion of its possible content in terms of component hardware, or at least its major assemblies. The first job is to set down that detail which can reasonably be relied upon to be present in the final design content. After this the gaps in the designer's knowledge must be filled in, using sensible assumptions.

In any product, the total cost of the unit is the sum of the costs of its component parts. It is extremely unlikely that all these components will be equal in cost. So they can be arranged in **probable cost order**, the most expensive at the top of the list, the cheapest at the bottom. This cost ordering will be based on the best information available to the designer at the time. Then, if we accept that the whole product cost is made up of a number of component costs arranged in descending order, a Pareto distribution is axiomatic.

In order to arrange components in descending cost order, the designer must first decide how many components there may be in the projected design. For a simple product this may present no problems. However, for a complex unit it may not be possible to predict accurately how many individual components there will be in the completed product. In this case, the best approach is to divide the unit into assemblies or major subassemblies and arrange these in probable cost order. Suppose we return to the vacuum cleaner.

There will be too many components for an accurate 'head count' this far ahead of detailed design, so we will consider only the major subassemblies which are likely to be present.

> main body assembly, motor assembly, front housing assembly, brush/beater bar assembly, front wheel assembly, rear wheel assembly, handle and flex assembly, storage bag assembly.

This is our best guess as to the probable subassembly content of the proposed design. Clearly, it is influenced by our personal knowledge of vacuum cleaners already on the market, so it is a well-informed guess. Figure 4.2 shows an exploded view of a typical vacuum cleaner, and this could form the basis of our predicted cost exercise.

It is fairly clear that the main body assembly with its wheel height adjuster, footswitch, bottom plate, and handle release mechanism, will be the most expensive of the subassemblies. Equally clearly, the rear wheel assembly is probably the cheapest.

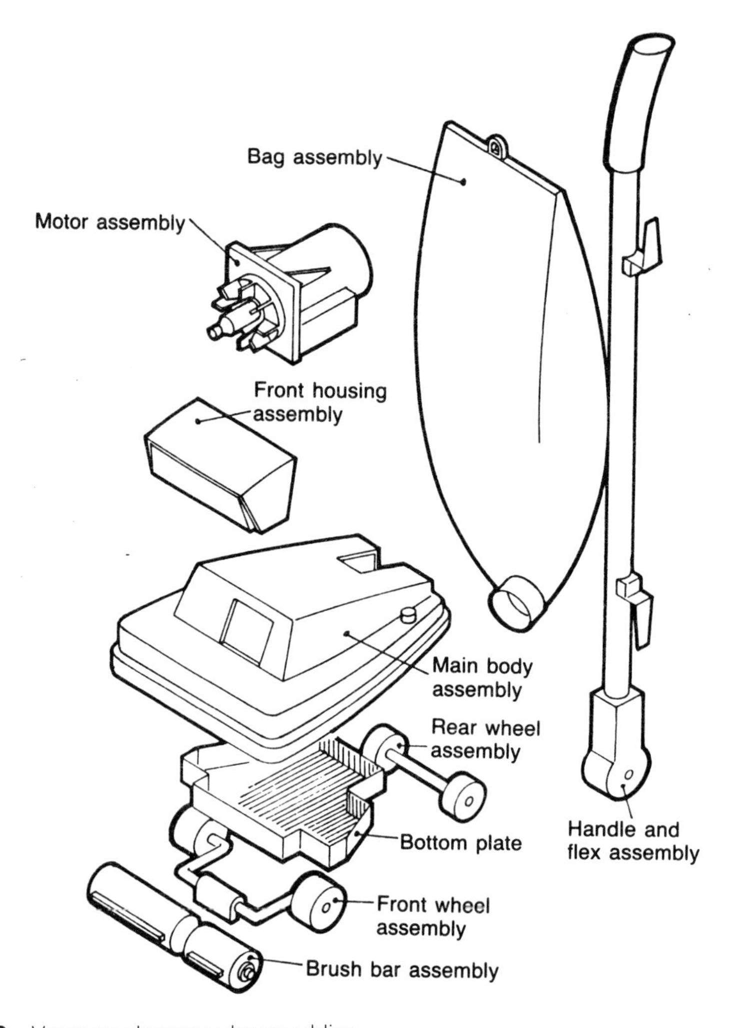

Figure 4.2 Vacuum cleaner subassemblies

Now we have to arrange the remaining subassemblies in probable cost order, between these two benchmarks.

main body assembly
motor assembly
handle and flex assembly
front housing assembly
bag assembly
brush bar assembly
front wheel assembly
rear wheel assembly

This listing is produced entirely from the best information available to the designer. The order of the assemblies within the listing is not critical; as more information is

Table 4.2 Pareto listing of subassemblies

Pareto	Subassembly	Cost fraction
0.75	main body	0.55
	motor	0.20
0.20	handle + flex	0.11
	front housing	0.05
	bag	0.04
0.05	brush bar	0.03
	front wheels	0.01
	rear wheels	0.01
1.00		1.00

generated during the design process, changes of order can be accommodated. What is important is that there should be no major items left out of the listing. Clearly, if the designer has background experience from previous vacuum cleaner designs, then his or her present knowledge will be extensive and the cost ordering will be fairly accurate. Also, if the designer's company is in the business of manufacturing vacuum cleaners, then actual costs of previous models will be available to help in the cost listing of the new unit.

Examination of the above cost listing might suggest that perhaps about 75% of the total cost is contained in the first two items, about 20% is in the next three items, and the balance of cost is in the remaining three items. Using this information, we can speculate as to how the cost is split between items. We know that they are in *descending* cost order, so each item must be less than, or equal to, that item which it follows (Table 4.2).

We have now established a rough relationship between costs of subassemblies, and if we can determine the probable actual cost of any one of them, the probable actual costs of the others can be calculated.

Suppose that from previous cleaner models we can set the cost of the motor assembly at around £4.00 (direct material + direct labour + overhead expense), then the cost of the complete cleaner is likely to be £4.00/0.20 = £20.00 (probable cost of subassembly divided by its cost fraction equals total unit cost).

From this, the cost of individual subassemblies can be assessed.

main body assembly cost is	£20.00 × 0.55 = £11.00
handle and flex assembly	£20.00 × 0.11 = £2.20
front housing assembly	£20.00 × 0.05 = £1.00 and so on.

To tidy up, we can now list the cleaner subassemblies, together with their probable actual costs, as shown in Table 4.3.

This is just a first stab at a probable cost, but it is not the whole picture. What we have done is to establish the probable cost of eight individual subassemblies. In order to make those subassemblies into a usable vacuum cleaner, they must be assembled, tested, packed and despatched. And we have not included the costs of these extra operations in our total. The actual cost of the motor assembly, £4.00, on which we based our costs, contained only material + direct labour + overhead expense.

Table 4.3 Probable costs of subassemblies

Subassembly	Cost fraction	Probable cost
main body	0.55	£11.00
motor	0.20	4.00
handle + flex	0.11	2.20
front housing	0.05	1.00
bag	0.04	0.80
brush bar	0.03	0.60
front wheels	0.01	0.20
rear wheels	0.01	0.20
totals	1.00	£20.00

Also our cost prediction is not based upon firm design drawings, so some of the information used is rather tentative. In addition, we may have produced this cost asessment without much previous estimating experience, and we should make allowance for this factor. To make our cost prediction more *professional*, we need to add a contingency.

assembly, testing, etc, add, say	10%
firmness of information, add, say	15%
estimating accuracy, add, say	15%
total contingency to be added	40%

so our total cost of £20.00 is multiplied by 1.4 to become £28.00 and this is called the ball-park manufacturing cost of the cleaner.

The cost of sale is found by adding amortised special-to-type expenses to the ball-park manufacturing cost. Let us assume that we have estimated amortised special-to-type expense at £2.00 per cleaner, sales and administration expense at 10% of cost of sale, and that we have set our profit margin at 40%. We can now proceed to predict further cost levels.

ball-park manufacturing cost	£28.00
amortised special-to-type expense	2.00
cost of sale	30.00
sales + administration expense	3.00
	33.00
profit margin (40% of above)	13.20
factory selling price	46.20
retailer mark up, say 30%	13.86
	60.06
VAT charged at 15%	9.00
probable price to customer	£69.06

Now we have a much clearer idea of the probable cost structure of the proposed new product and *we have not yet put pencil to paper to design it.* But we have made a number of important design decisions.

If the predicted cost is in line with our competition, the decision to launch the new product will be affirmative. The designer can now get on with the job of producing detailed designs for all the required components, and has a ready-made cost matrix within which to do this.

If, however, the predicted cost is too high, then the designer must look in detail at the cost compilation to see where savings can be made. This may entail downgrading the specification, using cheaper materials, cutting out any unnecessary features, reducing the labour content, etc., etc. In tackling the cost reduction exercise, it is necessary to start with those items at the top of the Pareto listing, i.e. those containing the greatest cost.

To make a saving of 50 pence, it is easier to make a 5% saving on the main body assembly, than to make a 50% saving on the front housing assembly. The former means a minor change, the latter would destroy the assembly.

Computer aided cost prediction

The foregoing explanation of the method of ball-park cost prediction has, of necessity, been fairly detailed. This has been done so as to illustrate the essential philosophy which is adopted in arriving very quickly at a reasonably accurate total cost, on which management may confidently base their future actions.

Clearly, as design proceeds there will emerge discrepancies between the components as originally envisaged, and those finally adopted. These discrepancies will be in both cost and performance. Usually such discrepancies result in the designer settling for more expensive components than those originally considered when making the cost prediction. Hence the reason for having some contingency. As the design progresses, some or all of this contingency will be absorbed.

Although the manual method of cost prediction described is not very time consuming with simple products, it can become tedious with large, multi-assembly products. So here, once again, the microcomputer can be used with advantage.

Essentially, the method entails committing to the computer memory details of three Pareto distributions, one high-knee, one medium-knee and one low-knee. This done, the operator must then input the following data.

which Pareto distribution, high, medium or low?
how many components/assemblies in the product?

The computer will then display a Pareto listing of cost factors, up to a maximum of eighty items. It will then ask for

the factor corresponding with the item of known cost
the cost of that item
the contingency factor
the amortised special-to-type expense
the sales and administration factor
the profit factor

```
50 PRINT CLS
100 PRINT"COMPUTER AIDED COST PREDICTION                    "
110 PRINT"______________________________            ________"
120 PRINT:PRINT
130 PRINT"ITEMS OF COST IN ANY UNIT ARE RELATED"
140 PRINT"BY A PARETO DISTRIBUTION WHICH MAY BE"
150 PRINT"HIGH, MEDIUM OR LOW KNEE IN FORM"
160 PRINT
200 DATA.095,.05,.025,.175,.085,.05,.24,.12,.075,.3,.15,.1,.35,.18
202 DATA.12,.4,.21,.14,.45,.24,.16,.485,.265,.18,.52,.29,.2,.555
204 DATA.315,.22,.58,.34,.24,.605,.365,.26,.63,.39,.28,.655,.415,.3
206 DATA.675,.44,.32,.695,.46,.34,.715,.48,.355,.73,.5,.37,.745,.52
208 DATA.385,.76,.54,.4,.775,.555,.415,.785,.57,.43,.795,.585,.445,.805
210 DATA.6,.46,.815,.615,.475,.825,.63,.49,.835,.645,.505,.843,.66,.52
212 DATA.85,.675,.535,.856,.69,.55,.862,.705,.565,.868,.72,.58,.874,.735
214 DATA.595,.88,.75,.61,.885,.765,.625,.89,.775,.64,.895,.785,.65,.9
216 DATA.795,.66,.905,.805,.67,.91,.815,.68,.915,.825,.69,.92,.835,.7
218 DATA.925,.845,.71,.93,.85,.72,.934,.855,.73,.938,.86,.74,.941,.865
220 DATA.75,.944,.87,.76,.947,.875,.77,.95,.88,.78,.953,.885,.788,.956
222 DATA.89,.796,.959,.895,.804,.962,.9,.812,.965,.905,.82,.968,.91,.828
224 DATA.97,.915,.836,.972,.92,.844,.974,.925,.852,.976,.93,.86,.978,.934
226 DATA.868,.98,.938,.876,.982,.942,.883,.984,.946,.89,.985,.95,.897,.986
228 DATA.954,.904,.987,.958,.911,.988,.962,.918,.989,.966,.925,.99,.97,.932
230 DATA.991,.974,.939,.992,.978,.946,.993,.982,.953,.994,.986,.96,.995,.99
232 DATA.967,.996,.993,.974,.997,.996,.981,.998,.998,.988,.999,.999,.995,1,1,1
240 DIM H(80),M(80),L(80),A(80),B(80)
250 FORJ=1TO80
260 READ H(J),M(J),L(J):NEXTJ
270 INPUT"WHICH DISTRIBUTION H,M,L ";A$
280 IFA$="H"THEN GOSUB1000
290 IFA$="M"THEN GOSUB2000
300 IFA$="L"THEN GOSUB3000
380 DIM F(A)
400 LET F(1)=B(1):PRINTF(1),TAB(11)"1"
410 FORJ=2TOA
420 LET F(J)=INT((B(J)-B(J-1))*1000+.5)/1000
430 PRINTF(J),J:NEXTJ
490 PRINT
500 INPUT"ITEM FRACTION ";C
505 PRINT
510 INPUT"ITEM COST ";D
520 E=D/C
525 PRINT
540 INPUT"CONTINGENCY FACTOR ";N
545 PRINT
550 P=INT((E*N)*100+.5)/100
560 PRINT"BALL PARK MANUFACTURING COST =";P
570 PRINT
580 INPUT"S T T EXPENSE AMORTISED ";S
590 PRINT
600 INPUT"SALES + ADMIN FACTOR ";SA
610 PRINT
620 INPUT"PROFIT FACTOR ";PF
630 PRINT
650 SP=INT(((P+S )*SA*PF)*100+.5)/100
660 PRINT"BALL PARK FACTORY SELLING PRICE =";SP
999 END
1000 PRINT:INPUT"HOW MANY ITEMS ";A
1002 IFA<1THEN1000
1004 IFA>80THEN1000
1010 FORJ=1TOA
1020 B(J)=H(INT((80/A)*J+.5)):NEXTJ:RETURN
2000 PRINT:INPUT"HOW MANY ITEMS ";A
2002 IFA<1THEN2000
2004 IFA>80THEN2000
2010 FORJ=1TOA
2020 B(J)=M(INT((80/A)*J+.5)):NEXTJ:RETURN
```

```
3000 PRINT:INPUT"HOW MANY ITEMS ";A
3002 IFA<1THEN3000
3004 IFA>80THEN3000
3010 FORJ=1TOA
3020 B(J)=L(INT((80/A)*J+.5)):NEXTJ:RETURN
READY.
```

Figure 4.3 Cost prediction program

It will then display the ball-park factory selling price. This is probably as far as the designer will need to go in assessing costs between alternative designs. But, of course, the program can be simply enhanced to include seller's mark up and VAT if required.

For the purpose of estimating the cost of a single component or subassembly, or a single prototype product, the same computer program may be used by inserting

zero for amortised special-to-type expense
one for sales and administration factor
one for profit factor.

The program and a sample printout are shown in Figures 4.3 and 4.4.

```
COMPUTER AIDED COST PREDICTION          CVS/1980
______________________________          ________

ITEMS OF COST IN ANY UNIT ARE RELATED
BY A PARETO DISTRIBUTION WHICH MAY BE
HIGH, MEDIUM OR LOW KNEE IN FORM

WHICH DISTRIBUTION H,M,L H
HOW MANY ITEMS  8
1            .555
 2            .205
 3            .096
 4            .054
 5            .04
 6            .026
 7            .014
 8            .01

ITEM FRACTION  .205

ITEM COST  4

CONTINGENCY FACTOR  1.4

BALL PARK MANUFACTURING COST = 27.32

S T T EXPENSE AMORTISED  2

SALES + ADMIN FACTOR  1.1

PROFIT FACTOR  1.4

BALL PARK FACTORY SELLING PRICE = 45.15
```

Figure 4.4 Printout of vacuum cleaner cost prediction

Costs of alternative manufacturing processes

At component level, when making the choice between alternative designs, the designer must consider all the principal attributes set out in the original model. And one of these most certainly will be cost.

As we have seen, component cost is made up of direct material plus direct labour plus overhead expense. When comparing alternative designs, it is important to consider each of these expenses in order to assess the differences in cost between alternative design strategies.

It is unlikely that there will be any really significant difference in overhead expense between two manufacturing departments in the same company, so this aspect of component cost may be ignored. On the other hand, material and labour costs may be quite different in alternative designs of the same component, so we should look at these in some detail.

There are two principal methods of material manipulation.

1 material cutting — which is roughly analagous to sculpture, chipping away the unwanted material in order to reveal the required shape in the raw material block.
2 material forming — which is analagous to pottery throwing, a doughy lump of raw material being pushed and pressed into the desired shape and size and finally set by the application of heat.

In general terms, material cutting involves starting with a block of material large enough to envelop the required component shape and size. The unwanted material is then cut away to release the component from its shroud. As much as 90% of the original block of material may be converted into chippings (swarf) by this process. The swarf so produced will recover only about 10% of its original cost when sold for scrap or recycling. So this form of manufacture usually incurs materials costs which are quite high in relationship to the material value in the finished component. On the plus side, it may be said that technology levels, tooling cost, and energy consumption are usually medium to low.

By comparison, there is very little material wastage in most material forming operations. The raw material usually comes in the form of powder or ingots, which are softened by heating and persuaded to flow into cavities of the required shape and size in appropriate tooling, sometimes under pressure. The only wastage is in runners, sprues and flash, and these are usually recoverable for re-use, although at some expense. On the other side of the coin, the material costs by weight or volume are generally medium to high; technology levels, tooling costs, and energy consumption are also much higher than for material cutting operations.

Summarising, we may say in general ...

material cutting	high material cost, medium to high labour costs, medium to low tooling and equipment costs, medium to low energy cost
material forming	medium material cost, low labour cost, high tooling and equipment costs, medium to high energy costs.

Keeping watch on material costs and how they change is not easy. Prices of raw

```
5 PRINT CLS
100 PRINT"MATERIAL PRICE OF BAR STOCK
110 PRINT"————————————————————————————
120 DATA1,1.12,2.2,6.8,7.7,12.11,12.8,14.32,16.5,5.3
130 DIMD(10),A$(10)
140 FORI=1TO10:READD(I):NEXTI
150 FORI=1TO10:READA$(I):NEXTI
160 DATA"BRIGHT MILD STEEL EN1A 220M07","40 CARBON STEEL EN8 080M40"
170 DATA"CAST IRON GRADE 20","ALUMINIUM H20","STAINLESS STEEL EN58A 302S25"
180 DATA"MANGANESE BRONZE","PHOSPHOR BRONZE","GUNMETAL","MONEL","BRASS"
190 PRINT
195 INPUT"TODAY'S PRICE OF BRIGHT MILD STEEL IN PENCE PER KILOGRAMME ";T
200 PRINT
210 FORI=1TO10:LETD(I)=INT((D(I)*T)*100+.5)/100:PRINTD(I),A$(I):NEXTI
READY.
```

Figure 4.5 Material price program

materials fluctuate widely and are dependent upon grade, quality, source and quantity. The method of material manipulation will also affect the price of the material in both its raw and finished states. The company purchasing department is a useful source of data and can usually supply details of current prices by weight, by volume, and by process, for the most commonly used engineering materials. The long-term trend of prices is upward, but the relative positions in the cost table seem to vary little over extended periods of time (reference 21).

A rough check on material prices can be had from a listing, related to a base material such as bright mild steel. Clearly, the contents of such a listing will depend upon the materials in which a particular designer is interested, and may be added to from time to time as material interests broaden. Such a listing, in the memory of a microcomputer, can be rapidly updated to provide a **first assessment** of probable alternative material costs when considering the early stages of a component design. The final choice between material alternatives will be made on more than cost information alone.

A very elementary material price program and printout are shown in Figures 4.5 and 4.6.

```
MATERIAL PRICE OF BAR STOCK
———————————————————————————

TODAY'S PRICE OF BRIGHT MILD STEEL IN PENCE PER KILOGRAMME
 40.5
 40.5              BRIGHT MILD STEEL EN1A 220M07
 45.36              40 CARBON STEEL EN8 080M40
 89.1              CAST IRON GRADE 20
 275.4              ALUMINIUM H20
 311.85              STAINLESS STEEL EN58A 302S25
 490.46              MANGANESE BRONZE
 518.4              PHOSPHOR BRONZE
 579.96              GUNMETAL
 668.25              MONEL
 214.65              BRASS
```

Figure 4.6 Material price comparison

Cost break points

It will be obvious that the cost of any component is largely determined by the processes which are used in its manufacture. The choice of manufacturing processes is governed predominantly by the quantities of components required.

Returning to the example of the simple spacing sleeve for the vacuum cleaner, let us see how it can be made. It can be manufactured by either material cutting or by material forming operations.

Consider firstly the material cutting strategy. It could be made ...

on a centre lathe	low tooling cost, long operation time, high labour skill, high labour cost
on a capstan lathe	medium tooling cost, medium operation time, low labour skill, low labour cost
on an automatic lathe	high tooling cost, very low operation time, nil direct labour cost

The total cost for manufacturing a particular quantity of spacing sleeves will be influenced by the setting-up cost, the component unit cost, and the quantity produced. This is shown graphically for the three alternative processes in Figure 4.7.

Key to Figure 4.7.

SL = set-up cost for centre lathe
SC = set-up cost for capstan lathe
SA = set-up cost for automatic lathe
UL = unit cost of components made on centre lathe
UC = unit cost of components made on capstan lathe
UA = unit cost of components made on automatic lathe
TL = total cost of components made on centre lathe
TC = total cost of components made on capstan lathe
TA = total cost of components made on automatic lathe
Q = batch quantity of components

For the centre lathe

$$TL = (SL/Q) + UL$$

for the capstan lathe

$$TC = (SC/Q) + UC$$

for the automatic lathe

$$TA = (SA/Q) + UA$$

A **cost break point** occurs where total component costs are equal by either process under consideration. For example, at quantity QLC at which the total component cost by centre lathe is equal to that by capstan lathe. At this quantity

$$(SC/QLC) + UC = (SL/QLC) + UL$$

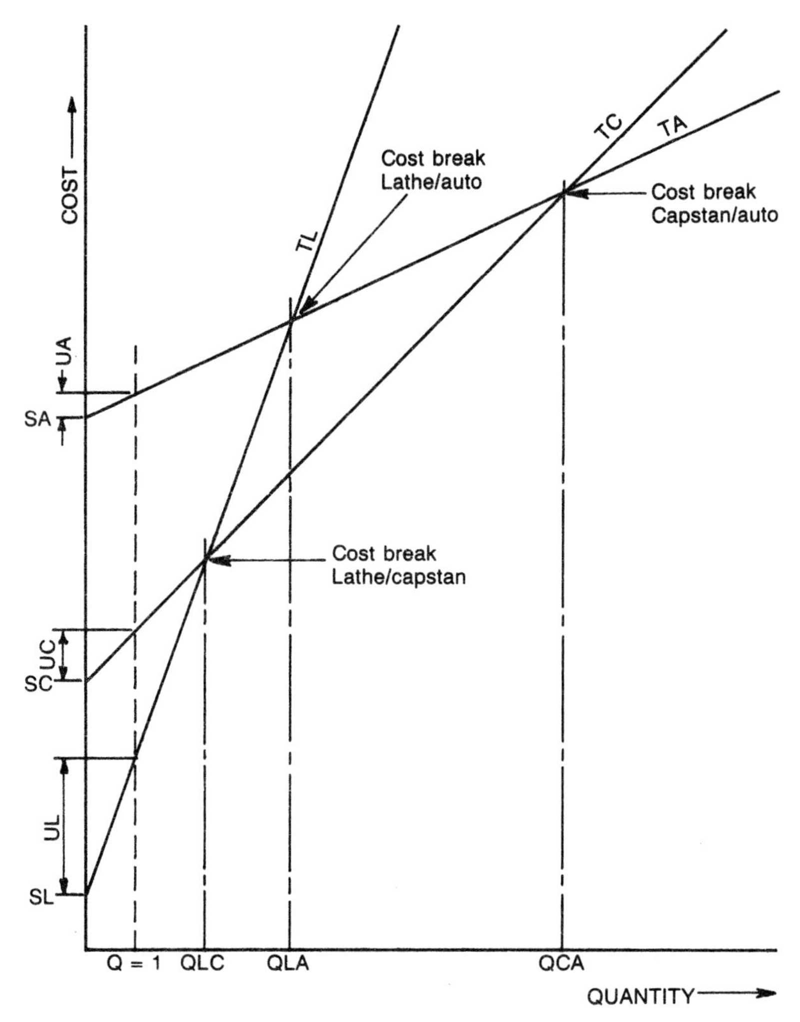

Figure 4.7 Cost break points

multiplying both sides of this equation by QLC

$$SC + QLC(UC) = SL + QLC(UL)$$

or $$SC - SL = QLC(UL - UC)$$

from which $$QLC = (SC - SL)/(UL - UC) \quad (1)$$

similarly $$QCA = (SA - SC)/(AC - UA) \quad (2)$$

and $$QLA = (SA - SL)/(UL - UA) \quad (3)$$

Thus, for quantities of components below QLC, it is cheaper to make them on the centre lathe; for quantities between QLC and QCA the capstan is the cheapest; for quantities above QCA the automatic lathe is cheapest. Consider secondly the material

```
5 PRINT CLS
100 PRINT"COST BREAK POINTS
110 PRINT"-----------------
120 PRINT
130 INPUT"NUMBER OF MANUFACTURING METHODS ";N
140 DIMA(N+1)
150 PRINT
160 FORI=1TON
170 PRINT"SET-UP COST ";I;:INPUTA(I):NEXTI
180 PRINT
190 FORI=1TON
200 PRINT"UNIT COST ";I;:INPUTB(I):NEXTI
210 PRINT
220 FORI=1TON-1
230 Q(I)=INT((A(I+1)-A(I))/(B(I)-B(I+1))*100+.5)/100
240 PRINT"BREAK EVEN QUANTITY ";I+1;",";I;" = ";INT(Q(I))
250 NEXTI
READY.
```

Figure 4.8 Cost break point program

```
COST BREAK POINTS

-----------------

NUMBER OF MANUFACTURING METHODS
 3

SET-UP COST  1  8
SET-UP COST  2  64
SET-UP COST  3  320

UNIT COST  1  2
UNIT COST  2  .27
UNIT COST  3  .01

BREAK EVEN QUANTITY  2 , 1  =  32
BREAK EVEN QUANTITY  3 , 2  =  984
```

Figure 4.9 Cost break point printout

forming strategy. It could be made in light alloy by...

impact extrusion plus a finishing operation	medium tooling cost, medium operation time, medium labour skill, medium labour cost
multicavity pressure die casting plus trimming	high tooling cost, very low operation time, medium labour skill, low labour cost

The quantity above which the casting process is cheaper is

$$QEC = (SC - SE)/(UE - UC)$$

i.e. the break-even point is the difference in set-up costs between casting and extrusion, divided by the difference in unit costs of the two processes. If a graph, similar to that in Figure 4.7, is constructed it will show the most economical method of manufacture for any batch quantity of components. A very simple computer program will also do this (Figures 4.8 and 4.9).

5 Standards and Standardisation

Some years ago the International Organisation for Standards (ISO) set up a committee to study the principles of standardisation, and the committee proposed the following definitions.

> Standardisation is the process of formulating and applying rules for an orderly approach to a specific activity for the benefit and with the cooperation of all concerned, and in particular for the promotion of overall economy, taking due account of functional conditions and safety requirements.
>
> A standard is the result of a particular standardisation effort, approved by a recognised authority. It may take the form of a document containing a set of conditions to be fulfilled; a fundamental unit or physical constant, e.g. ampere, absolute zero; or an object for physical comparison, e.g. metre.

Typical of international standards are the agreed sizes for engineering drawing sheets.

Designation	Size in mm
4A0	1682 × 2378
2A0	1189 × 1682
A0	841 × 1189
A1	594 × 841
A2	420 × 594
A3	297 × 420
A4	210 × 297
A5	148 × 210
A6	105 × 148

This range of sizes is based on a rectangle of one square metre in area, the sides of which are in the ratio of one to the square root of two. Individual sheet sizes are obtained by dividing the next larger size into two equal parts, the division being parallel to the shorter side, so that the areas of two successive sizes are in the ratio of two to one. The range A0 to A4 is considered adequate to most purposes.

British Standards Institution

The British Standards Institution (BSI) is responsible for the production of all national standards in the UK. It was formed in 1901 and was among the first organisations of its kind in the world. Its formation was the result of a need to bring some order into the rapidly expanding manufacturing industries. The standards issued by BSI may be grouped into five categories.

1 Those which establish standard terminology and symbols so as to provide a universally acceptable means of communication throughout industry.

2 Standard methods of testing which are essential to ensure that comparison of raw materials or finished products can be carried out on a uniform basis, so as to establish minimum acceptable levels of physical and mechanical properties, e.g. strength, hardness, viscosity, capacity, etc.

3 Dimensional standards which define form or shape. These are necessary to achieve interchangeability between mating parts and can also be used to define the range of sizes which are required to satisfy normal user needs.

4 Performance or quality standards specify the mechanical or electrical properties or performance of materials or finished components. They, in fact, ensure that the product will do the job expected of it. In many cases the standard specifies the final requirements, leaving the individual manufacturer to decide the method for meeting these requirements.

5 Codes of practice which set out methods of erection or of installation of equipment, or methods of operating or maintaining such equipment.

The request for a new standard can come from any responsible individual or group in industry, or it may originate from within BSI itself. The first action taken by BSI is to give the widest possible publicity to the project so that all interested parties may declare their views and, if necessary, participate in the drafting of the new standard.

Whilst the new standard is still in the draft stage, it is circulated amongst all sections of industry likely to be affected by it. Quite often this results in revision of the draft in order to make it completely acceptable to those who will later use it. This is most important in a country where a voluntary standards system operates. Compulsion, if any, comes from the customer. Similarly, most Government departments insist on compliance with relevant British Standards when placing contracts with industry.

The benefits of standardisation to the individual company, industry, and to the national economy are self-evident. But how is the designer affected? It may be considered by some that the only effect on the designer is to restrict his or her freedom of action. Although any form of standardisation must, by definition, be restrictive, this is a small price to pay for the advantages gained. Standards on such things as limits and fits, material specifications, processes, treatments, etc. actually help by relieving designers of the necessity for much investigation prior to making decisions. They can call for screwthreads, keyways, splines, gears, bearings, structural sections, and a whole range of standardised hardware, in the knowledge that these will have been produced to well proven standards, and that they will be entirely interchangeable with mating parts produced elsewhere to the same standards. Standards for drawing office practice ensure that the designer's own drawings will be correctly interpreted by engineers throughout industry, while other designers' drawings can be correctly interpreted. The designer knows that when he calls for an item to be manufactured to a national standard, the end product will be a well engineered component which is the result of considerable study by a number of very competent engineers.

Preferred numbers

Toward the end of the 19th century, a French balloonist and engineer Charles Renard was astounded to discover that 425 different sizes of cordage were in current use for mooring balloons. To reduce this unnecessary range, Renard devised a system of sizes which increased in steps, based on a geometric progression of weights per unit length of

Table 5.1 R series values

R5	R10	R20	R40	R40 theoretical
1.00	1.00	1.00	1.00	1.0000
			1.06	1.0593
		1.12	1.12	1.1220
			1.18	1.1885
	1.25	1.25	1.25	1.2589
			1.32	1.3335
		1.40	1.40	1.4125
			1.50	1.4962
1.60	1.60	1.60	1.60	1.5849
			1.70	1.6788
		1.80	1.80	1.7783
			1.90	1.8836
	2.00	2.00	2.00	1.9953
			2.12	2.1135
		2.24	2.24	2.2387
			2.36	2.3714
2.50	2.50	2.50	2.50	2.5119
			2.65	2.6607
		2.80	2.80	2.8184
			3.00	2.9854
	3.15	3.15	3.15	3.1623
			3.35	3.3497
		3.55	3.55	3.5481
			3.75	3.7584
4.00	4.00	4.00	4.00	3.9811
			4.25	4.2170
		4.50	4.50	4.4668
			4.75	4.7315
	5.00	5.00	5.00	5.0119
			5.30	5.3088
		5.60	5.60	5.6234
			6.00	5.5966
6.30	6.30	6.30	6.30	6.3096
			6.70	6.6834
		7.10	7.10	7.0795
			7.50	7.4989
	8.00	8.00	8.00	7.9433
			8.50	8.4140
		9.00	9.00	8.9125
			9.50	9.4406
10.00	10.00	10.00	10.00	10.0000

cordage. His scheme reduced the variety of mooring cordage to 17 sizes, and it was also applied to the rationalisation of pulleys, toggles and eyelets.

The range of numbers Renard used was the first practical application of preferred numbers.

The system of preferred numbers stems from the use of geometric rather than arithmetric progression. The ranges in use internationally today are designated R5, R10, R20, R40, etc., and they provide steps respectively of 5, 10, 20, 40, etc., over one order of magnitude, i.e. between 1 and 10, 10 and 100, 100 and 1000, etc.

The preferred number series is established by taking the first number of the series, multiplying it by a series factor, taking the resultant value and multiplying it by the same series factor, and so on until the whole series is built up.
These series factors are

> for the R5 series, the fifth root of 10, which is 1.5849
> for the R10 series, the tenth root of 10, which is 1.2589
> for the R20 series, the twentieth root of 10, which is 1.1220
> for the R40 series, the fortieth root of 10, which is 1.0593

The resulting numbers are rounded, by international agreement, to the most appropriate values. These are shown in Table 5.1.

The table shows the range from 1 to 10 for the basic series. Preferred numbers below 1 are found by dividing the values in Table 5.1 by 10, 100, etc., as required. Preferred numbers above 10 are found by multiplying the basic series by 10, 100, etc., as required.

A principal advantage to the designer of using preferred numbers is that they provide a guide which minimises unnecessary variations in size. They should in no way inhibit creativity by implying constraints on choice, but rather should assist the designer in avoiding proliferation in an arbitrary way. The adoption of preferred numbers in

```
5 PRINT CLS
100 PRINT"PREFERRED NUMBERS
110 PRINT"------------------
120 PRINT
130 DATA1,1.06,1.12,1.18,1.25,1.32,1.4,1.5,1.6,1.7,1.8,1.9,2,2.12,2.24,2.36
140 DATA2.5,2.65,2.8,3,3.15,3.35,3.55,3.75,4,4.25,4.5,4.75,5,5.3,5.6,6,6.3
150 DATA6.7,7.1,7.5,8,8.5,9,9.5,10
160 DIMA(41)
170 FORI=1TO41:READA(I):NEXTI
180 PRINT
190 INPUT"WHICH R SERIES 5,10,20,40? ";R
200 IFR=5THEN1000
210 IFR=10THEN2000
220 IFR=20THEN3000
230 FORI=1TO20:PRINTA(I):NEXTI
240 PRINT"PRESS ANY KEY TO CONTINUE"
250 GETA$:IFA$=""THEN250
260 FORI=21TO41:PRINTA(I):NEXTI:RESTORE
270 END
1000 FORI=1TO41STEP8:PRINTA(I):NEXTI:RESTORE:END
2000 FORI=1TO41STEP4:PRINTA(I):NEXTI:RESTORE:END
3000 FORI=1TO41STEP2:PRINTA(I):NEXTI:RESTORE:END
READY.
```

Figure 5.1 Preferred numbers program

```
PREFERRED NUMBERS


WHICH R SERIES 5,10,20,40?
 5
 1
 1.6
 2.5
 4
 6.3
 10
```

```
PREFERRED NUMBERS


WHICH R SERIES 5,10,20,40?
 10
 1
 1.25
 1.6
 2
 2.5
 3.15
 4
 5
 6.3
 8
 10
```

Figure 5.2 Specimen printouts of preferred numbers series

grading a range of articles, enables the required range to be covered by a minimum number of different sizes, with economic advantage to both manufacturer and user. Even if a range of sizes is not contemplated in the first instance, the selection of a preferred size for the initial design will facilitate the addition of further sizes should these be required subsequently.

A simple computer program (Figure 5.1) enables the designer to select at will any range of preferred numbers from the basic data in the R40 series, which is committed to the computer's memory. The program demonstrates simple manipulation of a database to produce a number of different outputs, and this technique can be applied equally well to more complex operations.

Tolerances, fits and limits

Whenever people set out to manufacture a product in quantity, they are always faced with a number of inconsistencies. For example

raw materials vary in their characteristics
people have differing skills and apply them variably
power supplies fluctuate
machine tools are not capable of exact repetition

climatic conditions are changeable
processes yield variable results
equipment is unreliable
inspection and test functions are subject to interpretation.

Variability

In face of these limitations it is impossible to produce economically a number of items having identical physical and performance characteristics. All human activities are subject to variability. But the fact that variability exists and can be measured, enables predictions to be made with a high level of confidence.

Normal frequency distribution

Variability tends to occur in a number of clearly recognisable patterns. The pattern which occurs most frequently in problems involving manufacturing process control is

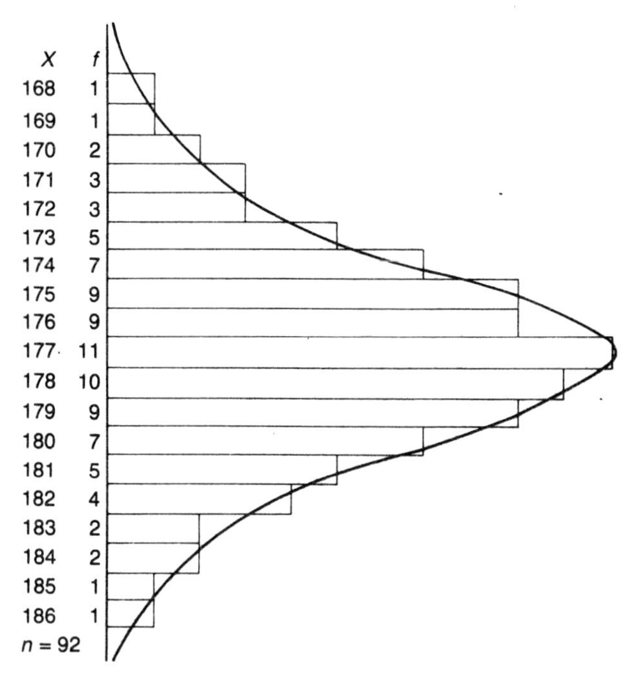

Figure 5.3 Histogram of male height measurements

the normal (Gaussian) frequency distribution. Let us examine a simple example of this normal frequency distribution.

If we take a reasonable cross-section of men from the population of a small community and record their individual measurements of height, we would expect to get a histogram as in Figure 5.3. The larger the sample taken from the original population, the nearer the histogram will conform to the actual distribution of male height in the community. The mean value of X is given by the summation of all values of X divided by the total number of occurrences.

$$\text{mean value } \bar{x} = \sum fX/n$$

X	f	fX
168	1	168
169	1	169
170	2	340
171	3	513
172	3	516
173	5	865
174	7	1218
175	9	1575
176	9	1584
177	11	1947
178	10	1780
179	9	1611
180	7	1260
181	5	905
182	4	728
183	2	366
184	2	368
185	1	185
186	1	186
totals	92	16284

$\bar{x} = 16284/92 = 177$ which is in centimetres.

The significance of any single occurrence is dependent upon its deviation (distance) from the mean value $\bar{x}$.

The average deviation for the distribution as a whole, is calculated by taking the root-mean-square of all deviations from the mean value. This is the **standard deviation** for the distribution, and it is denoted by the lower case Greek letter sigma σ.

$\sigma = \sqrt{\sum fd^2/n}$ where f = frequency of occurrence
d = deviation from mean value
n = total number of occurrences

Completing the tabulation ...

X	f	fX	d	d^2	fd^2
168	1	168	−9	81	81
169	1	169	−8	64	64
170	2	340	−7	49	98
171	3	513	−6	36	108
172	3	516	−5	25	75
173	5	865	−4	16	90
174	7	1218	−3	9	63
175	9	1575	−2	4	36
176	9	1584	−1	1	9
177	11	1947	0	0	0
178	10	1780	+1	1	10
179	9	1611	+2	4	36
180	7	1260	+3	9	63
181	5	905	+4	16	90
182	4	728	+5	25	100
183	2	366	+6	36	72
184	2	368	+7	49	98
185	1	185	+8	64	64
186	1	186	+9	81	81
totals	92	16284			1218

$\sigma = \sqrt{1218/92} = 3.64$ units of 1 cm = 3.64 cm

Knowing the characteristics of the normal frequency distribution curve, we can say with some certainty that the spread of all occurrences around the mean value will be

68.26% within ± one standard deviation
95.46% within $\pm 2\sigma$
99.73% within $\pm 3\sigma$

The 0.27% of occurrences which fall outside the 3σ limits take care of the occasional 2.5 metre giant and the 1.2 metre midget. Table 5.2 shows the approximate percentage of occurrences falling outside the specified limits for a normal frequency distribution.

Table 5.2 Percent occurrences outside standard deviations

1σ	31.74%
2σ	4.54%
3σ	0.27%
4σ	0.006%
5σ	0.00006%

Process control

Having shown how variability, although unavoidable, can be handled because its effects are predictable, let us look at a practical example which makes use of this

```
5 PRINT CLS
100 PRINT"MEAN & STANDARD DEVIATION
110 PRINT"------------------------
130 PRINT
140 INPUT"NUMBER OF VALUES ";N
150 DIMA(N)
160 PRINT
170 FORI=1TON:INPUT"NEXT VALUE ";A(I)
180 FX=FX+A(I)
190 M=FX/I:NEXTI
200 FORI=1TON:D2=(A(I)-M)↑2
210 FD=FD+D2
220 SD=SQR(FD/I):NEXTI
230 PRINT
240 PRINT"MEAN VALUE IS ";M
250 PRINT
260 PRINT"STANDARD DEVIATION IS ";INT(SD*100+.5)/100
READY.
```

Figure 5.4 Program for mean and standard deviation

knowledge. First, we set out a short computer program which will compute mean value and standard deviation from a series of values input to the computer (Figure 5.4).

Now suppose we are required to rough turn a number of steel blanks in a lathe to a diameter of 200 mm ± 3 mm. As the tolerance is very wide, we choose a well-used, old second-line lathe and proceed to run off the batch of components. During the first couple of hours work, we pick at random 20 machined blanks and measure their turned diameters, to the nearest millimetre, with the following result.

198	199	199	203	200	197	200	200	201	200
200	198	200	202	201	200	199	201	199	202

There are some interesting facts in these results.

all diameters measured are within the specified tolerance
we seem to be hitting the nominal dimension quite often
we are using the full range of the tolerance

At first glance, it appears the operation is well in control. But running the data through our little computer program produces the result shown in Figure 5.5. The mean value is almost exactly spot-on the nominal dimension, which is encouraging. As the frequency distribution appears to be normal, we would expect 99.73% of the machined components to fall within the range of $\pm 3\sigma$, that is

$$199.95 \text{ mm} \pm 4.29 \text{ mm}$$

But our tolerance requires

$$200 \text{ mm} \pm 3 \text{ mm}$$

Clearly, the old lathe is not capable of holding this tolerance, and we are totally out of control. If we continue with the machining, a significant portion of the finished batch will be above the top limit for diameter, and will have to be reworked at extra expense. Similarly, some of the items will be below the bottom limit for diameter and may have to be scrapped, at even greater expense, as they will have to be replaced by new components.

```
MEAN & STANDARD DEVIATION
________________________________

NUMBER OF VALUES?
 20

NEXT VALUE?  198
NEXT VALUE?  199
NEXT VALUE?  199
NEXT VALUE?  203
NEXT VALUE?  200
NEXT VALUE?  197
NEXT VALUE?  200
NEXT VALUE?  200
NEXT VALUE?  201
NEXT VALUE?  200
NEXT VALUE?  200
NEXT VALUE?  198
NEXT VALUE?  200
NEXT VALUE?  202
NEXT VALUE?  201
NEXT VALUE?  200
NEXT VALUE?  199
NEXT VALUE?  201
NEXT VALUE?  199
NEXT VALUE?  202

MEAN VALUE IS  199.95

STANDARD DEVIATION IS  1.43
```

Figure 5.5 Mean and standard deviation printout

All this assumes that a normal frequency distribution applies to the old lathe. It may well be that the distribution is **skewed** (7) toward one extreme or the other. In that case we may be faced with either more rework, or more scrap. It is a mistake for the designer to assume that every piece of plant in the machine shop is in tip-top condition and able to hold the tolerances claimed in the makers' specifications. Many of them never succeeded in doing so from the day they were installed, and age plays an important part in equipment accuracy.

In determining the tolerance to be applied to any dimension on a component, the designer must have in mind three things ...

the functional needs of the component in its working assembly
the capabilities of the manufacturing processes to be used
the means used to check if the specification has been met

Put more succinctly, the designer must recognise the compatibility between function, manufacture and inspection.

The fundamental basis of all tolerancing is the functional need of the product, and to regard every dimension and quality feature as a frequency distribution, is a good starting point.

Consider, for example, the brush bar of the vacuum cleaner. It is assembled on its spindle with two spacing sleeves as shown in Figure 5.6.

The length of the brush bar is 240 mm with a tolerance of ± 0.2 mm. The nominal

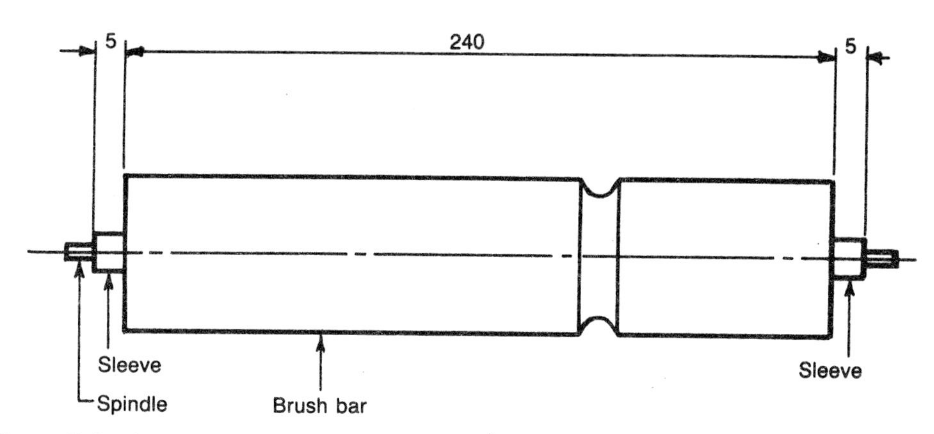

Figure 5.6 Cleaner brush bar assembly

length of each spacing sleeve is 5 mm. The whole assembly has to pass through an aperture of length 252 mm ± 0.6 mm, during the final assembly of the cleaner, with a clearance of from 1 to 3 mm.

Firstly, consider the case mathematically.

aperture size	252 ± 0.6
brush bar size	240 ± 0.2
room for two spacers	12 ± 0.8
nominal length of spacers	10 (tolerance not yet decided)
total tolerance available	2 ± 0.8

To achieve the desired clearance between the brush bar assembly and the aperture, we need a total tolerance of 2 mm ± 1 mm, i.e. 1 to 3 mm. So the total tolerance on the two spacers must be ±0.2 mm.

Each spacer size must be 5 mm ± 0.1 mm, brush bar size is 240 mm ± 0.2 mm, and the overall size of the brush bar assembly is

brush bar size	240 ± 0.2	
spacer number 1 size	5 ± 0.1	
spacer number 2 size	5 ± 0.1	
total assembly size	250 ± 0.4	
aperture size	252 ± 0.6	
total clearance	2 ± 1.0	as required.

However, the tolerances we have just set will impose their own cost levels on the three components of the brush bar assembly. If these tolerances could be relaxed, then the cost levels would be reduced and a cheaper brush bar assembly would result.

Secondly, consider the spacing sleeve as a component with variability. Its length has been set to a size of 5 mm ± 0.1 mm, and we may expect its actual length to vary in a

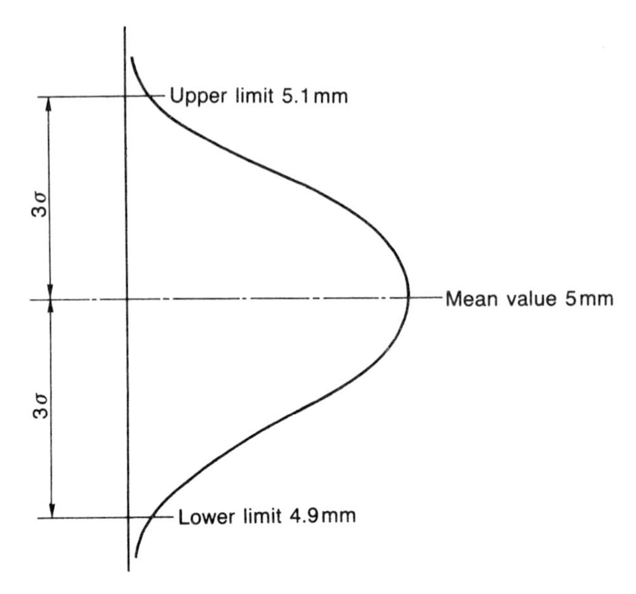

Figure 5.7 Spacer length variation

normal frequency distribution, as shown in Figure 5.7. Thus, in a box of 1000 spacers we would expect 99.73% to be within tolerance limits, i.e. 998 out of the 1000 would lie between 5.1 mm and 4.9 mm in length. So the chances of picking from the box one which was either on the top limit of 5.1 mm or on the bottom limit of 4.9 mm would be about one chance in 1000.

Similarly, the brush bar with a length of 240 mm ± 0.2 mm is also likely to vary in actual length in a normal frequency distribution, so that the chances of picking out a bar either on the top limit or the bottom limit is also about one chance in 1000.

Now the theorem of compound probabilities (7) states ...

> if a compound event be made up of a number of separate and independent sub-events, and the occurrence of the compound event be the result of each of these sub-events happening, the probability of the occurrence of the compound event is the product of the probabilities that each of the sub-events will happen.

So the chances of assembling a top limit brush bar with two top limit spacing sleeves is $1/1000 \times 1/1000 \times 1/1000$ or 1 chance in 1 000 000 000, and the odds are similarly minute of assembling three components which are all on the bottom limit.

The probabilities of combinations of parts can be assesed using the formula for the standard deviation of the sum of '*n*' independent variables.

$$\sigma_{\text{sum}} = \sqrt{(\sigma_{\text{part } 1})^2 + (\sigma_{\text{part } 2})^2 + \dots (\sigma_{\text{part } n})^2}$$

If the manufacturing processes of the brush bar and spacers produce lengths which vary within normal frequency distribution, then 99.73% of all components manufactured will be within $\pm 3\sigma$ of mean value.

brush bar tolerance is ± 0.2 mm $= 3\sigma$, so that $\sigma = 0.0666$ mm spacer tolerance is ± 0.1 mm $= 3\sigma$, so that $\sigma = 0.0333$ mm.

For the three component assembly we would expect

$$\sigma_{sum} = \sqrt{0.0333^2 + 0.0666^2 + 0.0333^2} = 0.0816 \text{ mm}$$

and as we would also expect the three component assembly to have an overall length which varies within $\pm 3\sigma$, the overall length of the assembly would be 250 mm $\pm$ 0.245 mm (i.e. 3×0.0816 mm).

As the total assembly size we can use is 250 mm $\pm$ 0.4 mm (page 000), we can afford to open up the tolerances on the three components by a factor of $0.4/0.245 = 1.63$ so that...

σ_{sum} becomes $0.0816 \times 1.63 = 0.133$ mm and three standard deviations, 3σ, for the assembly becomes $3 \times 0.133 = 0.40$ mm, which satisfies our overall tolerance requirement.

Thus the individual components can now be re-dimensioned.

brush bar	240 mm $\pm$ 0.32 mm
spacing sleeve	5 mm $\pm$ 0.16 mm

```
5 PRINT CLS
100 PRINT"MULTICOMPONENT ASSEMBLY
110 PRINT"―――――――――――――――――――――――
120 PRINT
130 INPUT"HOW MANY COMPONENTS IN ASSY ";N:PRINT
140 DIMA(N)
150 FORI=1TON:INPUT"TOLERANCE ";A(I):C=C+A(I)
160 S(I)=(A(I)/3)↑2
170 T=T+S(I):NEXTI
180 B=SQR(T)
190 R=C/(3*B)
200 PRINT:PRINT"COMPONENT TOLERANCES"
210 PRINT:FORI=1TON
220 PRINTI,INT((A(I)*R)*100+.5)/100
230 NEXTI
READY.
```

Figure 5.8 Multicomponent assembly program

```
MULTICOMPONENT ASSEMBLY
―――――――――――――――――――――――

HOW MANY COMPONENTS IN ASSY?
 3


TOLERANCE  .1
TOLERANCE  .2
TOLERANCE  .1

COMPONENT TOLERANCES
 1            .16
 2            .33
 3            .16
```

Figure 5.9 Printout for brush bar and spacer assembly

resulting in cheaper components which still meet the requirements of the assembly for size and for quality. These calculations can be handled by a short computer program, as in Figure 5.8.

Figure 5.9 shows a printout for spacing sleeves and brush bar assembly.

Worst case analysis

As has been shown, because of the variability present in all manufacturing operations, every component has to have physical tolerances. The effects of these tolerances on the finished components must be recognised by the designer when they are analysed for probable stresses.

For example, when designing a helical extension spring, it is necessary to calculate the maximum safe deflection permissible so that the spring will not be over-stressed in service. Allowing for Wahl's correction factor (4), the formula for maximum safe deflection contains three variables and is

$$\Delta = 0.0152\ ND^2/d$$

where Δ = maximum safe deflection in mm
N = number of active coils
D = mean coil diameter in mm
d = wire diameter in mm

Now we have already said that each dimension or quality of a component must have a manufacturing tolerance, and this applies to the dimensions of the spring. Let us apply reasonable tolerances.

Suppose $N = 20 \pm 0.5$ coils
$D = 8$ mm $\pm$ 0.8 mm
$d = 1$ mm $\pm$ 0.01 mm
material is carbon steel piano wire and the stress may not exceed 388 MN m^{-2}

Then if we completely ignore all manufacturing tolerances, the **nominal** safe deflection is

$$\Delta = 0.0152 \times 20 \times 64/1 = 19.45 \text{ mm}$$

This nominal value for maximum safe defection only applies **if ALL dimensions are on NOMINAL size,** a 1 in 1 000 000 000 chance. It is almost certain that the actual value for maximum safe deflection will be quite different from the nominal value. In the case of an extension spring, it is most important that the real maximum safe deflection is not exceeded, otherwise the spring will yield, take up a permanent set, and its characteristics will be altered rendering it useless for its designed job. So we must know **the worst case** which can apply and take precautions not to exceed this value.

To calculate Δ in the worst case we use the theory of variances, assuming that Δ is a function of N, D and d, and that there is no separate relationship between N, D and d. Using partial derivatives

$$\partial\Delta/\partial N = 0.0152\ D^2/d = 0.9728$$
$$\partial\Delta/\partial D = 0.0304\ ND/d = 4.864$$
$$\partial\Delta/\partial d = -0.0152\ ND^2/d^2 = -19.45$$

Table 5.3 Worst case analysis

No.	dN	dD	dd	dΔ	$\Delta = 19.45 + d\Delta$
1	+0.5	+0.8	+0.01	+4.1831	23.6331
2	−0.5	+0.8	+0.01	+3.2103	22.6603
3	+0.5	−0.8	+0.01	−3.5993	15.8507
4	−0.5	−0.8	+0.01	−4.5721	14.8779
5	+0.5	+0.8	−0.01	+4.5721	24.0221
6	−0.5	+0.8	−0.01	+3.5993	23.0493
7	+0.5	−0.8	−0.01	−3.2103	16.2397
8	−0.5	−0.8	−0.01	−4.1831	15.2669

The complete variance equation is

$$\begin{aligned} d\Delta &= (\partial\Delta/\partial N)\, dN + (\partial\Delta/\partial D)\, dD + (\partial\Delta/\partial d)\, dd \\ &= (0.9728)(\pm 0.5) + (4.864)(\pm 0.8) + (-19.45)(\pm 0.01) \end{aligned}$$

As we are dealing with maxima and minima for three variables, there are eight possibilities, $2^3 = 8$, as shown in Table 5.3. We are looking for the *MINIMUM* value of Δ and this occurs when the conditions of No. 4 above obtain. That is when the number of coils and the mean coil diameter are both on minimum sizes, and wire diameter is on maximum size. This value of 14.88 mm is 4.57 mm less than the calculated nominal value, i.e. an error of 23%. Had we been content to work to the nominal value, there would be a real risk of the spring being over-stressed in service.

But this is the arithmetical analysis. If we accept that the three variables are totally independent of each other and that each varies within a normal frequency distribution, then it is simple to calculate that the probable maximum safe deflection is 16.18 mm.

Fits and limits

When components are designed to be put together in an assembly, it is necessary for their sizes to be controlled so that a correct fit is achieved. Fits, which can vary from wide open clearance to heavy interference, are the subject of a British Standard Specification for ISO fits and limits published under the number BS 4500 : 1969 (20). This specification deals with all aspects of fits and limits and the reader is recommended to refer to it for a complete coverage of the subject.

For present purposes a simplified approach is attempted, and in order to avoid ambiguity, a number of definitions must be set out.

Basic size. The size by reference to which the limits of size are fixed. The basic size is the same for both members of a fit.

Deviation. The algebraical difference between a size and the corresponding basic size.

Upper deviation. The algebraical difference between the maximum limit of size and the corresponding basic size.

Lower deviation. The algebraical difference between the minimum limit of size and the corresponding basic size.

Tolerance. The difference between the maximum limit of size and the minimum limit of size, i.e. the algebraical difference between the upper deviation and the lower deviation. Tolerance is an absolute value without a sign.

Grade of tolerance. In a standardised system of limits and fits, a group of tolerances considered as corresponding to the same level of accuracy for all basic sizes.

Shaft. The term used by convention to designate all external features of a part, including parts which are not cylindrical.

Hole. The term used by convention to designate all internal features of a part, including parts which are not cylindrical.

Fit. The relationship resulting from the difference, before assembly, between the sizes of the two parts which are to be assembled.

Clearance fit. A fit which always provides a clearance, i.e. the tolerance zone of the hole is entirely above that of the shaft.

Interference fit. A fit which always provides an interference, i.e. the tolerance zone of the hole is entirely below that of the shaft.

Transition fit. A fit which provides either a clearance or an interference, i.e. the tolerance zones of the hole and the shaft overlap.

Shaft-basis system. A system of fits in which the different clearances and interferences are obtained by associating various holes with a single shaft. In the ISO system, the basic shaft is the shaft the upper deviation of which is zero.

Hole-basis system. A system of fits in which the different clearances and interferences are obtained by associating various shafts with a single hole. In the ISO system, the basic hole is the hole the lower deviation of which is zero.

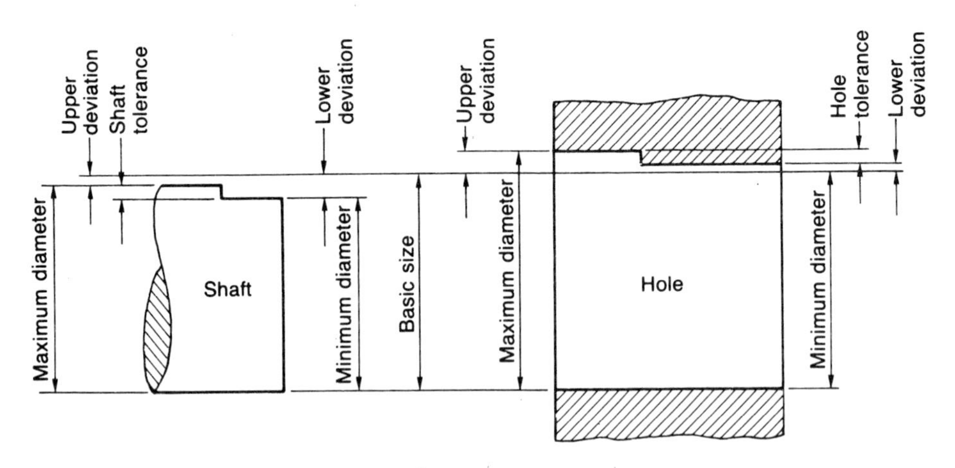

Figure 5.10 Features of hole and shaft fits

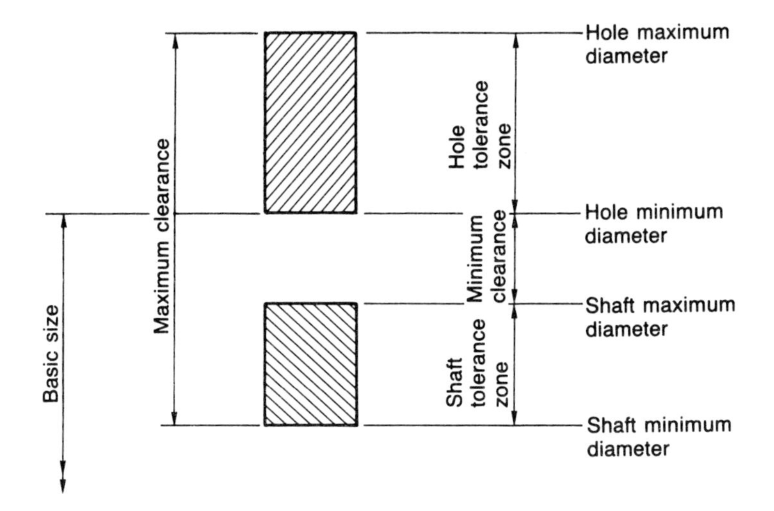

Figure 5.11 Conventional representation of clearance fit

Figure 5.11 shows the convention for a typical clearance fit condition in a hole-basis system. Each rectangle represents a tolerance zone; the upper for the hole, the lower for the shaft. In this case the hole tolerance zone is entirely above the shaft tolerance zone and, no matter what the sizes of the hole and the shaft within their respective tolerances, there will always be a positive clearance between them.

Consider the assembly of the vacuum cleaner brush bar spindle (the shaft) and the spacing sleeve (the hole). Assume the basic size of the shaft and the hole is 3 mm and that we require an **average** clearance. Reference to BS Data Sheet 4500A reveals six classes of clearance fit, which we will here designate...

sloppy
coarse
loose
average
close
precision

A fit of average clearance is designated H8/f7 and this fully describes the fit between the mating components.

Firstly, the hole. The letter H indicates a hole, the fundamental deviation of which is zero, regardless of the hole size. This means we are dealing with a basic hole, the lower limit of which is equal to the basic size of the fit, i.e. 3 mm.

The figure 8 indicates the tolerance grade, in this case 0.014 mm. Putting these data together gives a hole of dimensions 3.014/3.000 mm.

Secondly, the shaft. The letter *f* indicates a shaft which has an upper deviation from the basic size of −0.006 mm.

The figure 7 indicates a tolerance grade with a spread of 0.010 mm, i.e. a lower deviation of -0.016 mm, a shaft size of 2.994/2.984 mm. Extremes of tolerance on these mating components can now be calculated.

maximum tolerance = max hole − min shaft = 3.014 − 2.984 = 0.030 mm
minimum tolerance = min hole − max shaft = 3.000 − 2.994 = 0.006 mm

and both are positive showing that the hole size is always greater than the shaft size, i.e. a clearance fit.

If an interference fit had been required, reference to BS 4500A shows two such classes of fit, here designated light interference and heavy interference. Choosing a heavy interference for our two components, indicates a designation H7/s6, which gives the following conditions ...

hole size = 3.010/3.000 mm
shaft size = 3.020/3.014 mm
min tol = max hole − min shaft = 3.010 − 3.014 = −0.004 mm
max tol = min hole − max shaft = 3.000 − 3.020 = −0.020 mm

and both are negative showing that the hole size is always smaller then the shaft size, i.e. an interference fit.

The extremely wide choice of fits offered by the ISO system is far greater than required for most purposes, and the following selected range of fits is adequate for most applications.

sloppy clearance	H11/c11
coarse clearance	H9/d10
loose clearance	H9/e9
average clearance	H8/f7
close clearance	H7/g6
precision clearance	H7/h6
light transition	H7/k6
heavy transition	H7/n6
light interference	H7/p6
heavy interference	H7/s6

A fairly comprehensive computer program is needed to deal with fits and limits. It is shown in Figure 5.12 together with a printout for the brush bar spindle and spacing sleeves assembled with average clearance.

```
10 PRINT CLS
30 DIMH(4,23),E(10,23),X(10,23)
31 FORI=1TO4:FORJ=1TO23
32 READH(I,J):NEXTJ,I
50 PRINT"ISO FITS (BS4500)       C V STARKEY  4/82"
60 PRINT"__________________       ________________"
63 FORK=1TO10:FORJ=1TO23
64 READE(K,J):NEXTJ,K
70 PRINT"▒THE FOLLOWING TEN CLASSES OF FIT ONLY   ARE CONSIDERED:-▒"
71 FORK=1TO10:FORJ=1TO23
72 READX(K,J):NEXTJ,K
75 PRINT"1  SLOPPY CLEARANCE     H11C11"
76 PRINT"2  COARSE CLEARANCE     H9D10"
80 PRINT"3  LOOSE CLEARANCE      H9E9"
```

```
81 PRINT"4  AVERAGE CLEARANCE   H8F7"
82 PRINT"5  CLOSE CLEARANCE     H7G6"
83 PRINT"6  PRECISION CLEARANCE H7H6"
84 PRINT"7  LIGHT TRANSITION    H7K6"
85 PRINT"8  HEAVY TRANSITION    H7N6"
86 PRINT"9  LIGHT INTERFERENCE  H7P6"
87 PRINT"10 HEAVY INTERFERENCE  H7S6"
100 INPUT"▓GIVE NUMBER OF FIT ";A
105 IFA<1ORA>10THEN100
110 IFA=1THENI=1:K=1
120 IFA=2THENI=2:K=2
130 IFA=3THENI=2:K=3
140 IFA=4THENI=3:K=4
150 IFA=5THENI=4:K=5
160 IFA=6THENI=4:K=6
170 IFA=7THENI=4:K=7
180 IFA=8THENI=4:K=8
185 IFA=9THENI=4:K=9
190 IFA=10THENI=4:K=10
200 INPUT"▓NOMINAL BEARING SIZE ";B
210 IFB<=3THENJ=1
220 IFB>3ANDB<=6THENJ=2
230 IFB>6ANDB<=10THENJ=3
240 IFB>10ANDB<=18THENJ=4
250 IFB>18ANDB<=30THENJ=5
260 IFB>30ANDB<=40THENJ=6
270 IFB>40ANDB<=50THENJ=7
275 IFB>50ANDB<=65THENJ=8
280 IFB>65ANDB<=80THENJ=9
290 IFB>80ANDB<=100THENJ=10
300 IFB>100ANDB<=120THENJ=11
305 IFB>120ANDB<=140THENJ=12
310 IFB>140ANDB<=160THENJ=13
315 IFB>160ANDB<=180THENJ=14
320 IFB>180ANDB<=200THENJ=15
325 IFB>200ANDB<=225THENJ=16
330 IFB>225ANDB<=250THENJ=17
335 IFB>250ANDB<=280THENJ=18
340 IFB>280ANDB<=315THENJ=19
345 IFB>315ANDB<=355THENJ=20
350 IFB>355ANDB<=400THENJ=21
355 IFB>400ANDB<=450THENJ=22
360 IFB>450ANDB<=500THENJ=23
550 PRINT"▓HOLE DIMENSIONS ";B+H(I,J);"/";B
600 PRINT"▓SHAFT DIMENSIONS ";B+E(K,J);"/";B+X(K,J)
650 LETC=B+H(I,J)
660 LETD=B+X(K,J)
670 LETF=B+E(K,J)
700 V=INT((C-D)*1000+.5)/1000
710 W=INT((B-F)*1000+.5)/1000
720 PRINT"▓EXTREMES OF TOLERANCE ";V;"/";W
740 PRINT"▓"
750 INPUT"HARD COPY REQUIRED (Y/N) ";Z$
755 PRINT"▓▓"
760 IFZ$<>"Y"THEN 75
800 OPEN4,4
810 PRINT#4
820 PRINT#4,"NUMBER OF FIT ";A
830 PRINT#4:PRINT#4,"NOMINAL BEARING SIZE ";B
840 PRINT#4:PRINT#4,"HOLE DIMENSIONS ";C;"/";B
850 PRINT#4:PRINT#4,"SHAFT DIMENSIONS ";F;"/";D
860 PRINT#4:PRINT#4,"EXTREMES OF TOLERANCE ";V;"/";W
870 PRINT#4
880 CLOSE4
890 PRINT"▓▓"
900 GOTO75
999 END
```

(*continued*)

```
1980 REM H11
1990 DATA.06,.075,.09,.11,.13,.16,.16,.19,.19,.22,.22,.25,.25,.25,.29,.29,.29
1995 DATA.32,.32,.36,.36,.4,.4
2000 REM H9
2010 DATA.025,.03,.036,.043,.052,.062,.062,.074,.074,.087,.087,.1,.1,.1,.115
2015 DATA.115,.115,.13,.13,.14,.14,.155,.155
2020 REM H8
2030 DATA.014,.018,.022,.027,.033,.039,.039,.046,.046,.054,.054,.063,.063
2035 DATA.063,.072,.072,.072,.081,.081,.089,.089,.097,.097
2040 REM H7
2050 DATA.01,.012,.015,.018,.021,.025,.025,.03,.03,.035,.035,.04,.04,.04
2055 DATA.046,.046,.046,.052,.052,.057,.057,.063,.063
2060 REM C11
2070 DATA-.06,-.07,-.08,-.095,-.11,-.12,-.13,-.14,-.15,-.17,-.18,-.2,-.21
2075 DATA-.23,-.24,-.26,-.28,-.3,-.33,-.36,-.4,-.44,-.48
2080 REM D10
2090 DATA-.02,-.03,-.04,-.05,-.065,-.08,-.08,-.1,-.1,-.12,-.12,-.145,-.145
2095 DATA-.145,-.17,-.17,-.17,-.19,-.19,-.21,-.21,-.23,-.23
2100 REM E9
2110 DATA-.014,-.02,-.025,-.032,-.04,-.05,-.05,-.06,-.06,-.072,-.072,-.084
2115 DATA-.084,-.084,-.1,-.1,-.1,-.11,-.11,-.125,-.125,-.135,-.135
2120 REM F7
2130 DATA-.006,-.01,-.013,-.016,-.02,-.025,-.025,-.03,-.03,-.036,-.036,-.043
2135 DATA-.043,-.043,-.05,-.05,-.05,-.056,-.056,-.062,-.062,-.068,-.068
2140 REM G6
2150 DATA-.002,-.004,-.005,-.006,-.007,-.009,-.009,-.01,-.01,-.012,-.012,-.014
2155 DATA-.014,-.014,-.015,-.015,-.015,-.017,-.017,-.018,-.018,-.02,-.02
2160 REM H6
2170 DATA0,0,0,0,0,0,0,0,0,0,0,0,0,0,0,0,0,0,0,0,0,0,0
2180 REM K6
2190 DATA.006,.009,.01,.012,.015,.018,.018,.021,.021,.025,.025,.028,.028
2195 DATA.028,.033,.033,.033,.036,.036,.04,.04,.045,.045
2200 REM N6
2210 DATA.01,.016,.019,.023,.028,.033,.033,.039,.039,.045,.045,.052,.052,.052
2215 DATA.06,.06,.06,.066,.066,.073,.073,.08,.08
2220 REM P6
2230 DATA.012,.02,.024,.029,.035,.042,.042,.051,.051,.059,.059,.068,.068,.068
2235 DATA.079,.079,.079,.088,.088,.098,.098,.108,.108
2240 REM S6
2250 DATA.02,.027,.032,.039,.048,.059,.059,.072,.078,.093,.101,.117,.125,.133
2255 DATA.151,.159,.169,.19,.202,.226,.244,.272,.292
2260 REM X11
2270 DATA-.12,-.145,-.17,-.205,-.24,-.28,-.29,-.33,-.34,-.39,-.4,-.45,-.46,-.48
2275 DATA-.53,-.55,-.57,-.62,-.65,-.72,-.76,-.84,-.88
2280 REM X10
2290 DATA-.06,-.078,-.098,-.12,-.149,-.18,-.18,-.22,-.22,-.26,-.26,-.305,-.305
2295 DATA-.305,-.355,-.355,-.355,-.4,-.4,-.44,-.44,-.48,-.48
2300 REM X9
2310 DATA-.039,-.05,-.061,-.075,-.092,-.112,-.112,-.134,-.134,-.159,-.159,-.185
2315 DATA-.185,-.185,-.215,-.215,-.215,-.24,-.24,-.265,-.265,-.29,-.29
2320 REM X7
2330 DATA-.016,-.022,-.028,-.034,-.041,-.05,-.05,-.06,-.06,-.071,-.071,-.083
2335 DATA-.083,-.083,-.096,-.096,-.096,-.108,-.108,-.119,-.119,-.131,-.131
2340 REM XG6
2350 DATA-.008,-.012,-.014,-.017,-.02,-.025,-.025,-.029,-.029,-.034,-.034
2355 DATA-.039,-.039,-.039,-.044,-.044,-.044,-.049,-.049,-.054,-.054,-.06,-.06
2360 REM XH6
2370 DATA-.006,-.008,-.009,-.011,-.013,-.016,-.016,-.019,-.019,-.022,-.022
2375 DATA-.025,-.025,-.025,-.029,-.029,-.029,-.032,-.032,-.036,-.036,-.04,-.04
2380 REM XK6
2390 DATA0,.001,.001,.001,.002,.002,.002,.002,.002,.003,.003,.003,.003,.003
2395 DATA.004,.004,.004,.004,.004,.004,.004,.005,.005
2400 REM XN6
2410 DATA.004,.008,.01,.012,.015,.017,.017,.02,.02,.023,.023,.027,.027,.027
2415 DATA.031,.031,.031,.034,.034,.037,.037,.04,.04
2420 REM XP6
2430 DATA.006,.012,.015,.018,.022,.026,.026,.032,.032,.037,.037,.043,.043,.043
```

```
 2435 DATA.05,.05,.05,.056,.056,.062,.062,.068,.068
 2440 REM XS6
 2450 DATA.014,.019,.023,.028,.035,.043,.043,.053,.059,.071,.079,.092,.1,.108
 2455 DATA.122,.13,.14,.158,.17,.19,.208,.232,.252
READY.
```

NOTE: in lines 70, 100, 200, etc., reverse field Q = cursor down.

```
NUMBER OF FIT  4

NOMINAL BEARING SIZE  3

HOLE DIMENSIONS  3.014 / 3

SHAFT DIMENSIONS  2.994 / 2.984

EXTREMES OF TOLERANCE  .03 / 6E-03
```

Figure 5.12 Program with printout for fits and limits

Part 2
GUIDED DESIGN ASSIGNMENT

This assignment will show the evolution of product design from basic requirements to final general proposals. The objective will be the design of a range of small-power electric motors, formerly called fractional-horsepower motors, to be manufactured in significant numbers over a five year period.

To simplify the understanding of the design process, one particular size of motor in the middle of the range will be used as a demonstration vehicle.

But first, some essential background to the project. What is a small-power motor, sometimes called a fractional-horsepower motor, or just FHP? British Standard BS5000 : Part 11 : 1973 (17) defines it thus:

> '... a.c., d.c. and universal rotating electrical machines of any continuous rated output up to and including 0.75 kW or kVA per 1000 rev/min (synchronous speed for induction machines), and rated voltages up to 250 V d.c. or single-phase a.c. ...'

The importance of small-power motors in modern life cannot be over-emphasised. They are in universal use in industry, commerce and the home. Some typical applications ...

> small machine tools such as lathes, drills, saws, etc.; mixers; compressors; pumps; conveyors; stackers; moulding machines; automatic controls; etc.
>
> business machines; fans; blowers; compressors; mixers; printers; cold stores; computer peripherals; air conditioning plant; dry cleaning plant; shredders; slicers; pumps; typewriters; sorters; collaters; processing equipment; etc.
>
> washing machines; refrigerators; hair driers; vacuum cleaners; food mixers; freezers; tumble driers; lawn mowers; central heating installations; DIY power tools; record and cassette players; timers; clocks; etc.

6 Introduction and Background

The description **small-power** covers a variety of motor types but, in this assignment, we shall be concerned only with one type, the induction motor. The term **induction** indicates a machine in which the rotating element, the **rotor** has currents flowing in it which are **induced** by currents flowing in the stationary element, the **stator** through the action of interlinking magnetic fields.

There is no physical connection between the supply voltage and the rotor, which is usually of the **squirrel-cage** type, so-called because of its geometrical similarity to the rotating cages used for exercising pet squirrels, mice, and hamsters.

Unlike polyphase motors, single-phase induction motors are not self-starting and require external help to get them rotating. Once started, however, they continue to rotate unaided and develop useful power. The different types of induction motors are usually described by the arrangements which are employed to get them started, thus ...

split-phase induction-run; capacitor-start induction-run; capacitor-start capacitor-run; repulsion-start induction-run; shaded-pole.

In this assignment we will consider a range of four variants, designated by power output and sub-divided by shaft speed. The power outputs will be as recommended in BS5000, i.e. 120 W, 180 W, 250 W and 370 W.

This selection is marginally different from the equivalent preferred sizes in the R20 series which are 125, 180, 250 and 355. It is, however, a fairly close approximation to the former fractional-horsepower ratings for 1/6 HP(124 W), 1/4 HP(186 W), 1/3 HP(248 W) and 1/2 HP(373 W). It is probably also a good compromise which allows direct replacement of older motors by newer varieties.

In order to better understand the features of the motor we are about to design, it is necessary to look at some of the technical characteristics of two of the motor varieties, the split-phase and the capacitor-start types (6, 15). The reason for looking at these two types is because we may want to combine certain aspects of both in the one design, so that standardisation of motor frame size and shaft size can be achieved, with resultant cost savings.

Split-phase induction-run motor

Split-phase induction-run motors have a run winding and a start winding, with their axes displaced 90 electrical degrees, connected as shown in Figure 6.1.

The start winding has a higher resistance to reactance ratio than the run winding, so the two currents are not in phase, as is shown in the vector diagram. Since the start winding current I_s leads the run winding current I_r, the stator magnetic field first

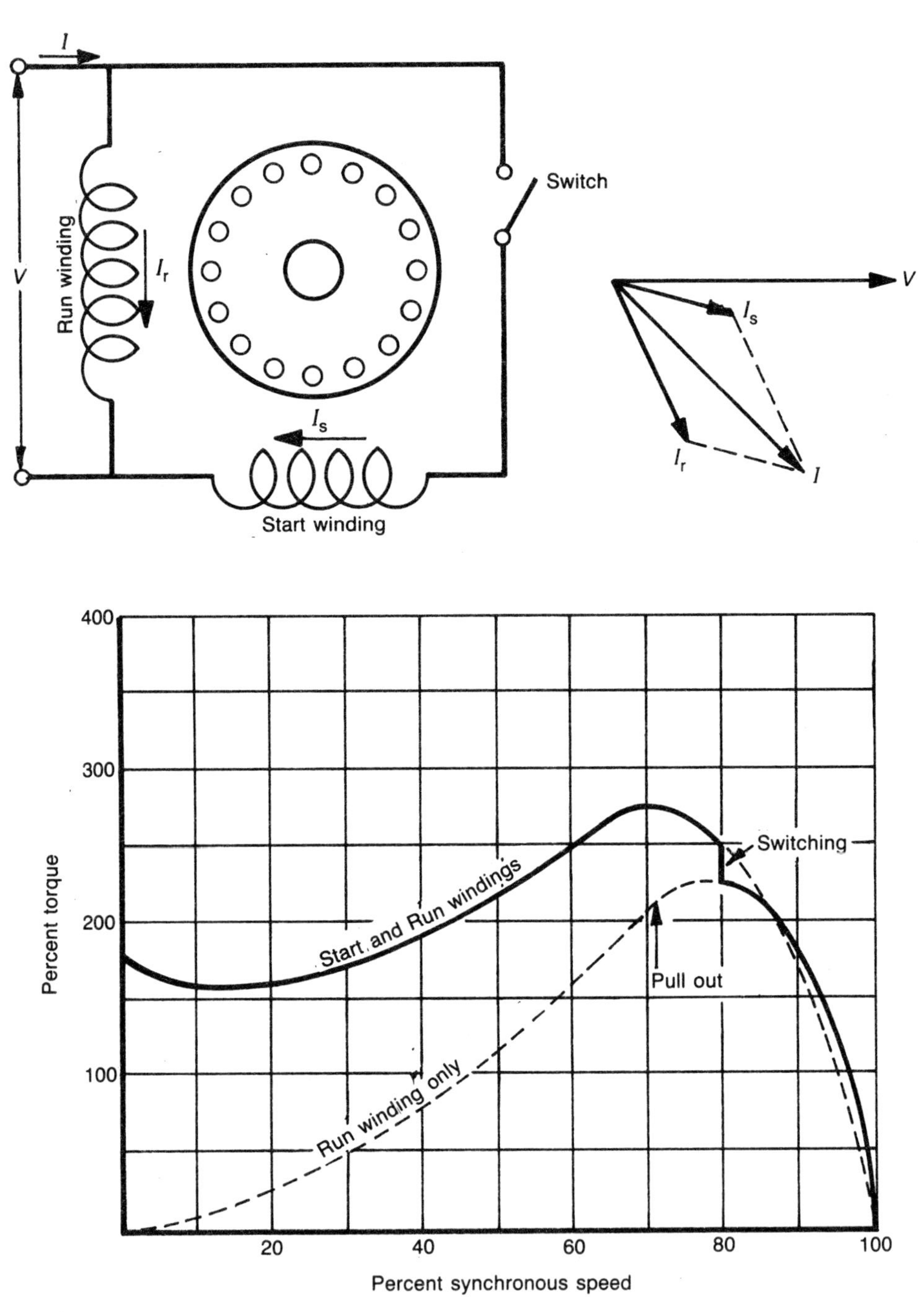

Figure 6.1 Split-phase induction motor

reaches a maximum along the axis of the start winding and then, later in time, reaches a maximum along the axis of the run winding. The winding currents are the equivalent of two unbalanced 2-phase currents, and the motor acts as an unbalanced 2-phase machine. The result is a rotating magnetic field which causes the motor to start rotating.

Starting torque is between 150% and 200% of full-load torque, and reaches a maximum of almost 300% during run-up. At about 80% of synchronous speed, the start winding is disconnected automatically by a switch, and run-up to full speed is completed on the run winding only. Synchronous speeds are 3000, 1500 and 1000 rev/min, for 2-, 4-, and 6-pole machines respectively. Actual full-load speeds are about 95% of these values, i.e. 2850, 1425, and 950 rev/min.

Capacitor-start induction-run motor

The capacitor-start induction-run motor (Figure 6.2) is essentially a split-phase machine in which the time displacement between the two winding currents is increased by means of a capacitor in series with the start winding. As with the normal split-phase type, the start winding is disconnected automatically at about 80% of synchronous speed, thus both winding and capacitor are in use for very short, intermittent periods.

Dry type electrolytic capacitors are normally used for this class of motor, as they are intermittently rated and inexpensive. By use of an appropriate capacitor, the start winding current I_s can be made to lead the run winding current I_r by 90 electrical degrees, as it would in a balanced 2-phase machine. However, the best compromise between starting torque, starting current, and cost produces a phase angle of something less than 90 degrees. A typical torque-speed curve is shown in Figure 6.2, featuring the very high starting torque available from this type of motor. With both windings in circuit, starting torque is around 300% to 400% of full-load torque, and this rises marginally during run-up. These motors are used where high starting torque is essential, e.g. for pumps, compressors, etc., and also in areas where supply voltage may fluctuate downwards from normal level.

In both types of motor, split-phase and capacitor-start, the high resistance to reactance ratio for the start winding is achieved by using a smaller diameter wire than that used for the run winding. About half the cross-sectional area is usual. Although windings with small diameter wire tend to overheat quickly, this use is permissible as the start winding is in circuit only intermittently. This is during start-up, or when the motor speed drops through overload below the pull-out torque-speed of the run winding, which is around 70% of synchronous speed. When this happens, the start winding is automatically switched back into circuit to assist the run winding to deal with the overload. Excessive operation of the start winding in overload conditions quickly raises the temperature of the start winding and, at about 50°C above ambient, a built-in thermal overload device will operate to switch off the supply current and protect the winding from damage.

Having sketched in some background, we may now turn our attention to the design of the range of small-power motors.

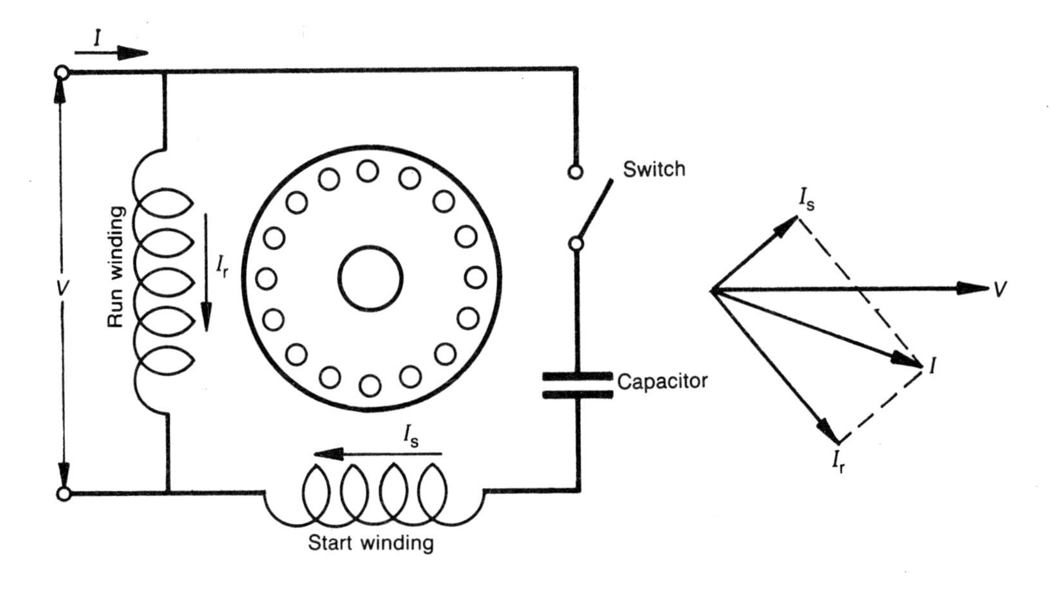

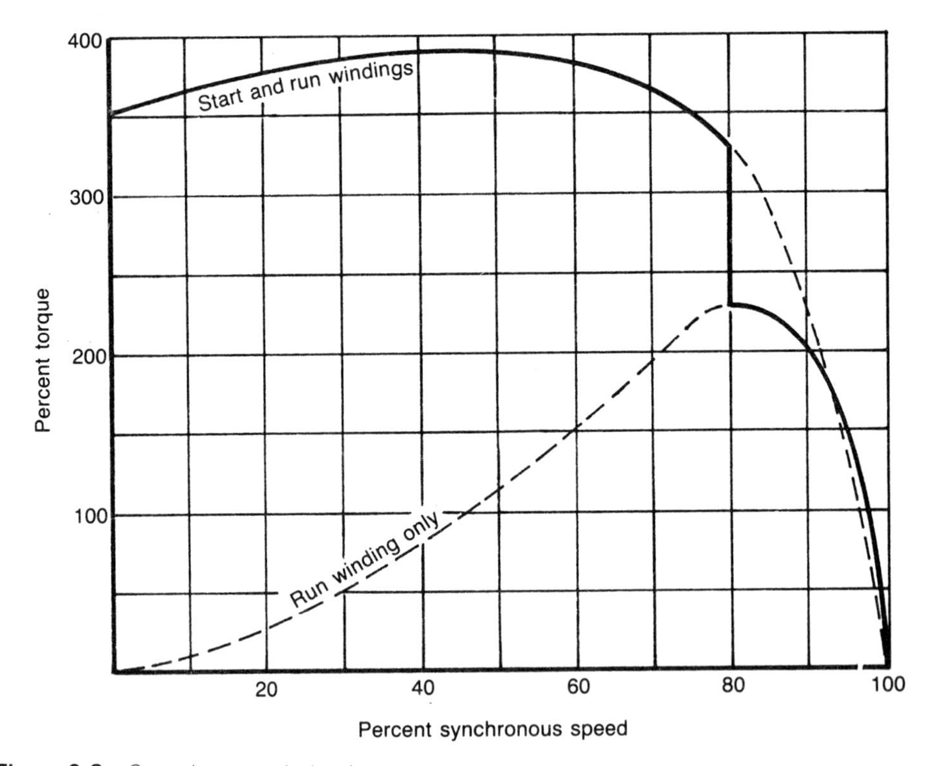

Figure 6.2 Capacitor-start induction motor

Problem finding

This exercise will already have been carried out at top management level and its results can be summarised thus ...

Need	supplement Company product range with portable power units for customer equipment drives	
Problem	produce motors (design range of small-power motors for manufacture and sale)	
Parameters	Labour	skilled and semi-skilled male and female with medium level technology
	Materials	general engineering, i.e. steel, plastics, copper, aluminium, etc., with medium-level technology equipment
	Time	introduce within 12 months
	Space	available factory area in square metres
	Money	£500 000 for product introduction.

This sets the scene in general terms and we must now proceed with the creative phases of the project.

Problem solving

The designer will receive from management a fairly comprehensive brief, indicating Company requirements for the proposed range of the new product. This will include details of the types of motor envisaged, the quantities of each variant to be made over the manufacturing period, the spread of special-to-type expenditure, the principal competitors' types, and any other relevant data which may help to establish a firm base from which the new design will be developed. Note! For reasons of simplicity, the present assignment cannot include all these data, nor will every possible alternative be looked into.

The following pages are meant to highlight the general approach by a designer to a real commercial situation, showing the gradual growth of knowledge about the new product as progress is made toward the ultimate Company objectives. So let us proceed.

The range of motors to be considered will be 120W, 180W, 250W, and 370W variants, in 2-, 4-, and 6-pole types. The quantities envisaged are 20 000, 35 000, 50 000, 50 000, 50 000, in each of the five years of planned manufacture.

year	120W	180W	250W	370W	totals
1	2800	8000	5600	3600	20000
2	4900	14000	9800	6300	35000
3	7000	20000	14000	9000	50000
4	7000	20000	14000	9000	50000
5	7000	20000	14000	9000	50000
totals	28700	82000	57400	36900	205000

This programme might break down into yearly loadings as shown in Table 6.1.

Table 6.1 Proposed manufacturing programme

year 1

variant	120W	180W	250W	370W	totals
2-pole	1000	2000	2000	1600	6600
4-pole	1000	3000	2000	1200	7200
6-pole	800	3000	1600	800	6200
totals	2800	8000	5600	3600	20000

year 2

variant	120W	180W	250W	370W	totals
2-pole	2000	3000	3000	2000	10000
4-pole	2000	7000	4800	3000	16800
6-pole	900	4000	2000	1300	8200
totals	4900	14000	9800	6300	35000

years 3, 4 and 5

variant	120W	180W	250W	370W	totals
2-pole	2500	6000	5000	2000	15500
4-pole	4000	9000	7000	6000	26000
6-pole	500	5000	2000	1000	8500
totals	7000	20000	14000	9000	50000

Table 6.1 forms the basis for our manufacturing programme for the five year period and at this point we have enough information to summarise our activities by variant. These data will be useful when we attempt the rationalisation of motor shaft sizes, and are contained in Table 6.2.

Table 6.2 Manufacturing programme by variant

years 1 to 5

variant	120W	180W	250W	370W	totals
2-pole	10500	23000	20000	9600	63100
4-pole	15000	37000	27800	22200	102000
6-pole	3200	22000	9600	5100	39900
totals	28700	82000	57400	36900	205000

Initial cost prediction

At the earliest moment we must try to establish a probable cost for the product. To do this we utilise all the information currently at our disposal.

Special-to-type expense

From experience of projects of a similar nature, we believe the special-to-type expense for the motors will be around £70 000 to £80 000 made up roughly as follows...

tooling	£50 000
test gear (use general purpose equipment)	nil
technical labour (1 engineer for 1 year)	£12 000
excess costs (labour and materials)	£10 000
total special-to-type expenses	£72 000

The expenses will be amortised (recovered) over the first full year of manufacture, i.e.

$$£72\,000/20\,000 \text{ units} = £3.60 \text{ per unit.}$$

Product cost

Reference to Figure 6.3 will show the content of a typical motor as manufactured by our competitors. It shows a sleeve-bearing split-phase variant, whereas we will be designing a ball-bearing version. However, although our proposed design when developed may differ in some details, it is unlikely to be radically changed in its presentation.

So we may with some confidence base our cost prediction on the contents of Figure 6.3. A probable list of major components is

rotor and shaft assembly
drive frame assembly
casing
stator assembly
start and run windings
motor foot
terminal frame assembly
ball bearings (2 off)
start winding switch assembly
tie bolts (4 off)

A first stab at putting these items into descending cost order might look like this

rotor and shaft assembly
terminal frame assembly
drive frame assembly
start and run windings
start winding switch assembly
stator assembly
casing
ball bearings (2 off)
tie bolts (4 off)
motor foot

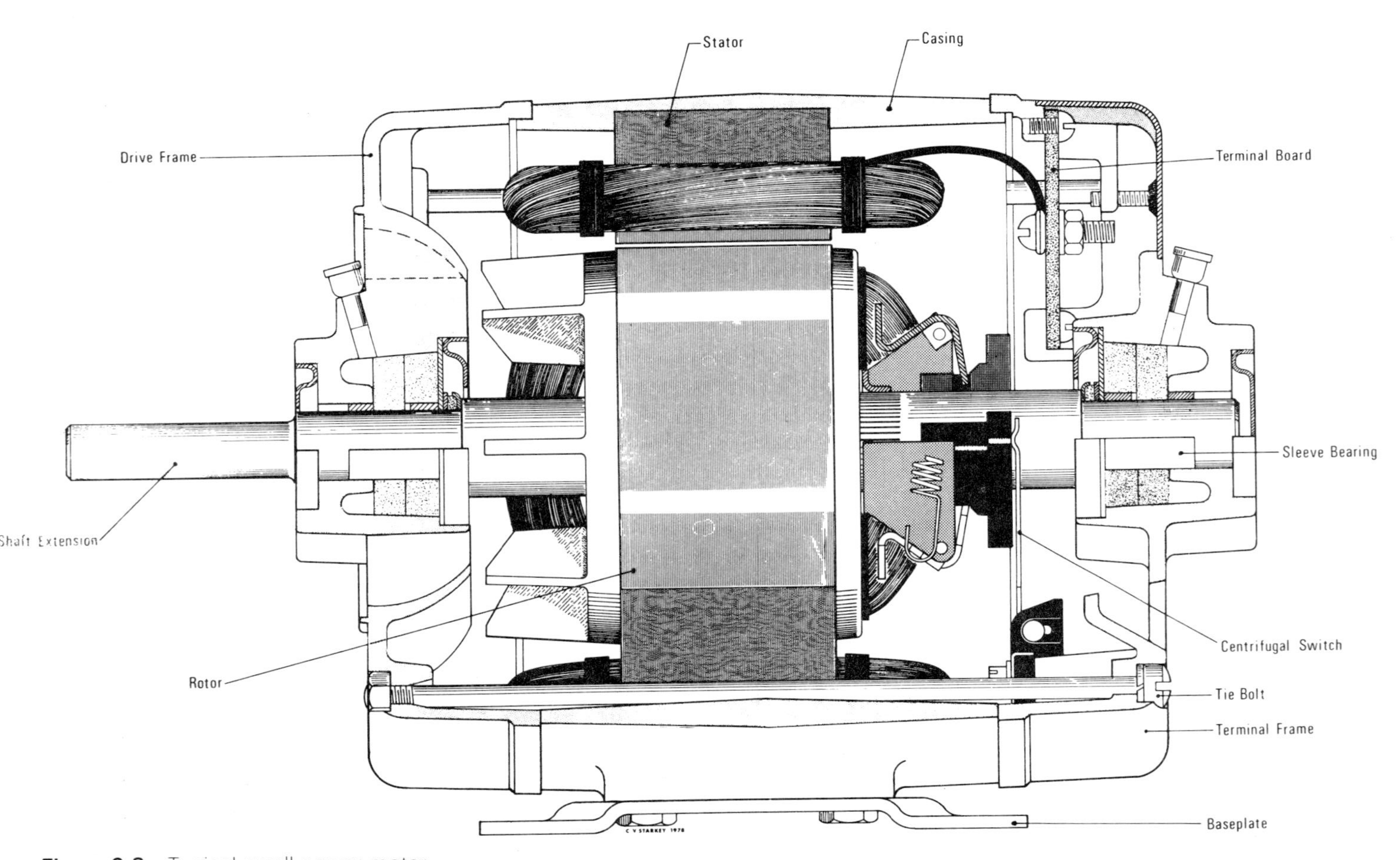

Figure 6.3 Typical small-power motor

About the only proprietary article in this listing is the pair of ball bearings. They are rather far down the listing for use as a basis for accurate cost prediction. However, they are the only item on which we can gain factual cost evidence, so we will use them. This will give us our first idea of product cost, which we can then update progressively as firmer information becomes available.

As yet we have no firm idea on which to base our choice of bearing size, so an informed guess will have to suffice. Inspection of the competitive motor may suggest a bearing of around 20 mm bore, in a 'light series' range, so we will settle for this as a starting point. Enquiry of a bearing stockist reveals a probable cost of £0.75 each in quantities of 20 000.

Before we can produce a cost prediction, we need to decide the following ...

which Pareto curve shall we use, high-, medium- or low-knee?
amount of contingency for missing items, information firmness, accuracy of estimating?
special-to-type expense amortisation?
sales and administration expenses?
company profit margin?

```
COMPUTER AIDED COST PREDICTION


ITEMS OF COST IN ANY UNIT ARE RELATED
BY A PARETO DISTRIBUTION WHICH MAY BE
HIGH, MEDIUM OR LOW KNEE IN FORM

WHICH DISTRIBUTION H,M,L L
HOW MANY ITEMS  10
1           .18
 2            .16
 3            .12
 4            .12
 5            .1
 6            .08
 7            .068
 8            .062
 9            .056
  10            .054

ITEM FRACTION  .062

ITEM COST  1.5

CONTINGENCY FACTOR  1.2

BALL PARK MANUFACTURING COST = 29.03

S T T EXPENSE AMORTISED  3.6

SALES + ADMIN FACTOR  1.1

PROFIT FACTOR  1.15

BALL PARK FACTORY SELLING PRICE = 41.28
```

Figure 6.4 Initial cost assessment

By inspection of the competitive motor we can see that there are no individual items of outstandingly high cost in relationship to any other item, so a low-knee Pareto curve would seem to be appropriate.

For contingency purposes, a 5% allowance for combined assembly, test, and miscellaneous small items of hardware would be about right. As the information we have is reasonably firm (we do have the motor of our competitor in front of us) a 5% allowance will suffice. For estimating accuracy we should allow perhaps 10% because the starting point, the cost of bearings, is quite low in relationship to overall motor cost. This gives an overall contingency of 20%.

Special-to-type expense amortisation has already been fixed at a figure of £3.60 per unit.

Sales and administration expense will be set at 10% of cost of sale. Company profit for this class of product is 15%.

As we would normally sell the product direct to the user, the actual price paid by the customer will include VAT, currently 15%. Let us use our computer program to arrive at an initial cost assessment (Figure 6.4).

Customer buying price will be around £47.47 including VAT. This is accurate enough for a decision to be made about the price competitiveness of the product. Assuming the decison is affirmative, we can proceed with the design, and update the cost prediction, en route, as further information is generated.

The motor frame

We now face the first fundamental decison. What size shall we make the motor frame? This decision is not simple. In fact it is highly complex. We have a number of alternatives, each with its advantages and its disadvantages.

The motor frame can vary from being small in diameter and long, **sausage-like**, to being large in diameter and short, **pancake-like**.

The characteristics of the **sausage** are
- small diameter, robust laminations
- relatively inexpensive lamination press-tools
- small capacity power presses for lamination manufacture
- very difficult coil-winding with small-bore stator stack
- most winding wire is in active turns, hence good motor performance
- poor cooling features, hence poor motor performance.

The characteristics of the **pancake** are
- large diameter laminations susceptible to damage
- very expensive lamination press-tools because of size
- large capacity power presses required for manufacture
- very easy coil-winding, large diameter short stack
- most winding wire is in end turns, hence poor motor performance
- good cooling features, hence good motor performance.

However, the market survey which has decided our five year manufacturing plan, will have identified the motor type required as being **square-like**, in which the rotor stack diameter approximates to its length. On this basis we can proceed to decide the most probable physical sizes of the motor frame.

Author's note. In the real world, quite successful sausage-like and pancake-like versions of small-power motors have been developed and marketed, usually for very specific duty requirements. However, in the context of a design course of study of perhaps only 60 hours duration, it is necessary to limit the choices open to the student artificially.

Stator size

The volume of active material (iron) in a stator stack is generally related to the power required from the motor. In determining the stator size, the most important dimension is the outer diameter D which is made as large as possible within the motor frame. Selection of the outer diameter is usually from historical records of previous motors, either own-make or those of competitors.

Under the Imperial system of measurement, six inches was a popular stator diameter and many successful designs were based upon this size. It would be sensible to use this figure as a starting point for the present design. D will be 15.2 cm and this is a convenient size for punching from 16 cm (preferred size) strip, while leaving enough 'webbing' along each edge of the strip for feeding through a progression press tool. To determine the size of our mid-range motor, 180 W 4-pole machine, we use the chart in Figure 6.5.

The value of D^2L for a 180 W 4-pole machine is 1150. If we divide this value by 15.2^2 we get

$$L = 1150/231.04 = 4.977 \text{ cm, say 5 cm.}$$

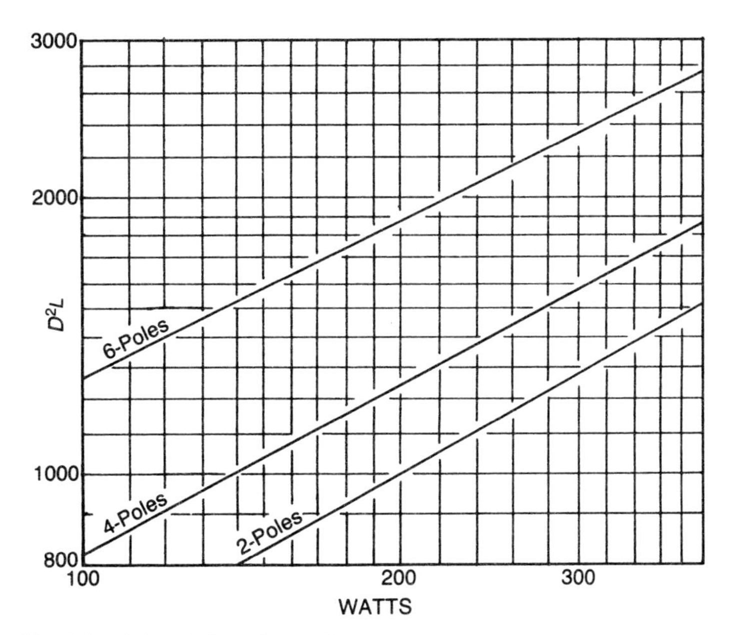

Figure 6.5 Chart for determining stator size

So we have a stack length L of 5 cm for our mid-range machine. Our design will grow around a stator size of 152 mm diameter and 50 mm in length. *THE FIRST FUNDAMENTAL DECISION HAS BEEN SUCCESSFULLY MADE.*

The ratio of stator bore to stator outer diameter has been determined empirically over very many successful designs, and is taken as 0.59 for 4-pole machines.

This gives a bore size of $15.2 \times 0.59 = 8.97$ cm, say 9 cm, and we call this $D1$.

Rotor size

The radial airgap between the stator bore $D1$ and the rotor diameter $D2$, for bores less than 20 cm, is given by the empirical formula

$$G = 0.127 + 0.0042\ D1/P^{1/2}$$

where P is the number of poles
G and $D1$ are in millimetres

Thus, $$G = 0.127 + 0.0042(90)/2 = 0.32 \text{ mm}$$

and rotor diameter $$D2 = D1 - 2G = 90 - 0.64 = 89.36 \text{ mm.}$$

Thus, the basic size of the rotor stack will be 89.36 mm diameter and its length will be the same as the stator stack, i.e. 50 mm. This rotor stack will be mounted on a shaft which, in turn, will be mounted on two ball bearings. To decide the approximate positions for the two ball bearings, we must first consider what other features of the design have to be placed between the rotor stack and the bearings. Looking back to Figure 6.3, we have

the ventilation fan at the drive end
the start winding switch assembly and terminal board at the terminal end.

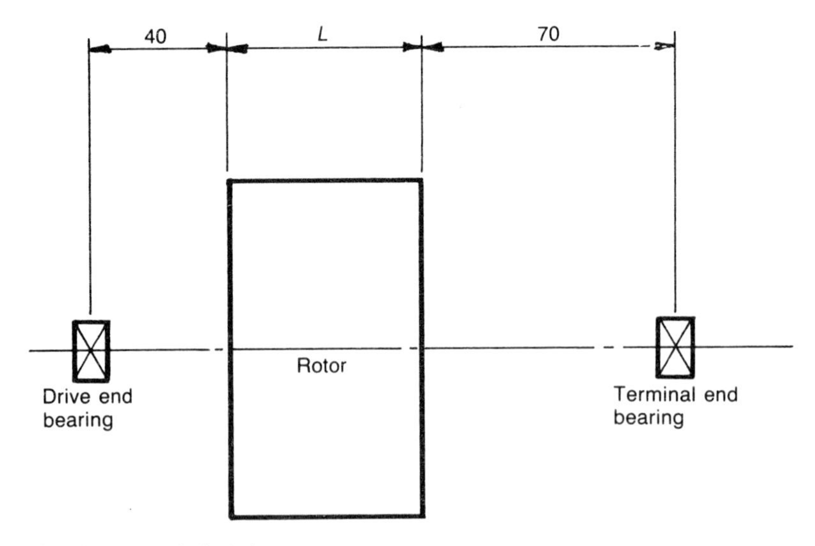

Figure 6.6 Provisional shaft layout

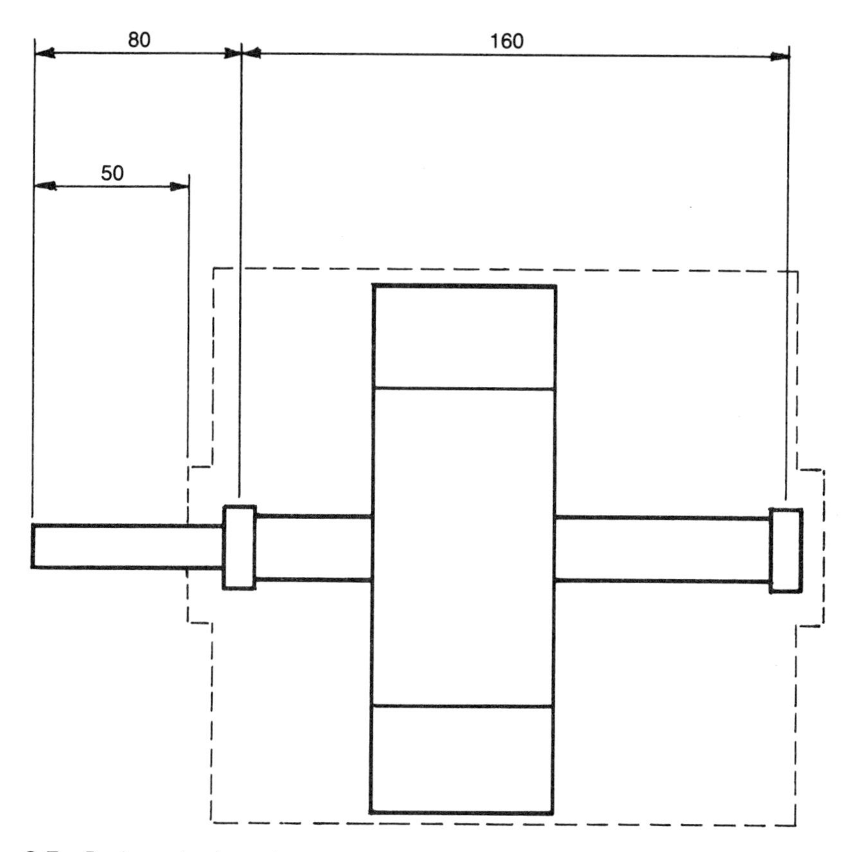

Figure 6.7 Basic motor layout

Making provisional allowances for these items, within our present knowledge, will establish approximate positions for the ball bearings as shown in Figure 6.6.

If we now allow, say, 50 mm of shaft extension beyond the drive end frame, the approximate basic sizes of the motor emerge as shown in the sketch Figure 6.7.

So, in addition to our first fundamental decision, we have just made a number of quite significant intermediate decisions. None of these should be regarded as inviolate. As design proceeds, it may well be necessary to change our ideas and to modify some of the dimensions which we have just set. But, for the time being, we have created some data on which we can proceed to build our motor design.

Attribute model

We must now consider the principal attributes to be incorporated into our new product. Again, past experience and contact with customers is the most reliable method of fixing these. If the designer could arrange to accompany a service engineer on visits to motor users, an enormous amount of useful information about end-user needs would be gathered. For the present exercise, let us utilise such information as may have been derived from such visits, and select the following principal attributes ...

appearance
cost
interchangeability
life
maintainability
performance
reliability
safety

Once again these have been listed alphabetically so that no preference is shown. The criteria which sell small-power motors are primarily performance and reliability, and these are almost equal in importance to the average end-user. A fit-and-forget product which is safe and inexpensive, easy to maintain, does its job consistently and reliably forever, just about sums up most customers' view of the ideal motor. So let us run some figures through our computer program and establish a reasonable attribute model.

	Comparisons						
Attribute	1/2	1/3	1/4	1/5	1/6	1/7	1/8
appearance	25	35	30	28	20	22	24
cost	75						
interchangeability		65					
life			70				
maintainability				72			
performance					80		
reliability						78	
safety							76

These figures yield the following attribute model ...

0.19 performance
0.17 reliability
0.15 safety
0.14 cost
0.12 maintainability
0.11 life
0.09 interchangeability
0.05 appearance

which presents us with reasonable guidelines as we set out on the initial design exercise. Due importance seems to have been settled in the attributes of performance and reliability as required, and the basic sizes of the motor have already been tentatively settled; now these bones must be clothed.

We start by examining the required outputs and the available inputs.

outputs: 180 watts at 1425 rev/min
inputs: 250 volts 50 Hz.

These outputs will be delivered via the motor shaft extension, which will transmit power to the equipment to be motorised. The possible transmission methods might be

belt drive
chain drive
gear drive
gearbox
coaxial coupling

In the first three modes, the motor shaft extension will be subjected to torsion and to bending. In the last two modes, it will experience torsion only. Clearly, the first three modes represent more onerous operating conditions, so these should be considered in detail (worst case), and the others should be ignored.

The designer has no way of knowing exactly how the motor will be used in service, so some reasonable assumptions must be made. We might assume that in the case of belt, chain or gear drive, the effective diameter of the pulley, sprocket or pinion would be about 50 mm.

Reserve factor

When a motor is switched on, the application of torque is **sudden**, i.e. there is no slow build-up of load. The effect of sudden load application is to double both the stress and deflection experienced by the shaft, and this must be catered for in the designer's calculations. Motor shaft material, historically, is free-cutting mild steel. This is a general-purpose material with good mechanical strength and easy-machining properties.

Determination of the actual size of the cross-section of a component is usually dependent upon its required strength in service. This, in turn, is related to the physical properties of the material from which it is made. The property of most interest to the designer is the ultimate strength of the selected material. This represents a measure of the stress which can be absorbed by the material before the onset of failure. Clearly, if components are designed so that in service they are stressed right up to the ultimate strength of the parent material, then enormous numbers of failures will occur. These failures will be due to factors over which no control can be exercised, such as ...

indiscernable and unavoidable imperfections in material and workmanship
material deterioration due to atmospheric or electrolytic corrosion
unforeseen possibility of changes in the amount or mode of loading.

Thus, the designer must choose, almost arbitrarily, a value of stress significantly lower than the ultimate strength of the material, beyond which the component will not be taken. The relationship between this **safe working stress** and the ultimate strength, can be called the **reserve factor**.

In setting the safe working stress for a material for use in a component, the designer must have in mind four areas of criticality.

1 The relationship of yield stress or proof stress to the ultimate strength, which may be known or can be found from reference books, or from tests (4, 5). As a guide, the following may be used for steels...
   0.43 for austenitic chromium–nickel steels
   0.50 for mild steels and some un-heat-treated carbon steels
   0.57 for 1% nickel steels in the normalised condition
   0.67 for manganese–nickel–molybdenum steels
   0.75 for 1% nickel steels in hardened and tempered condition
   0.76 for 1.5% nickel–chromium–molybdenum steels
   0.80 for 2.5% nickel–chromium–molybdenum, and martensitic chromium–nickel steels
   Note! It should be emphasised that these values may vary depending upon the condition in which the material is supplied and also on its limiting ruling section (18, 19).
2 An ignorance or uncertainty factor which must make allowance for accidental overload, unreliable material quality, corrosive attack, poor surface finish, etc.
3 Stress concentration due to the presence of **stress raisers** such as undercuts, keyways, splines, shoulders, screwthreads, etc.
4 How the load is to be applied, i.e. gradually, suddenly, with impact, or whether it will fluctuate and cause premature failure through fatigue.

Items 1, 3 and 4 above can usually be verified from reference data or calculated, so the **guesstimation** will normally be confined to item 2 only. The reserve factor finally selected will be the lowest value resulting from the four items above. The intention is to select a safe working stress which is no higher than the lowest stress which may give rise to premature failure, through whatever cause.

In the case of the motor shaft, we must recognise the possibility of fatigue due to rapid reversals of loading. The following illustration shows a generalised fatigue curve for steel subjected to rotating bending loads, as would arise in fatigue testing. From this it is clear that we should set the safe working stress, in item 4 above, no higher than 0.5 of the ultimate strength of the material (Figure 6.8).

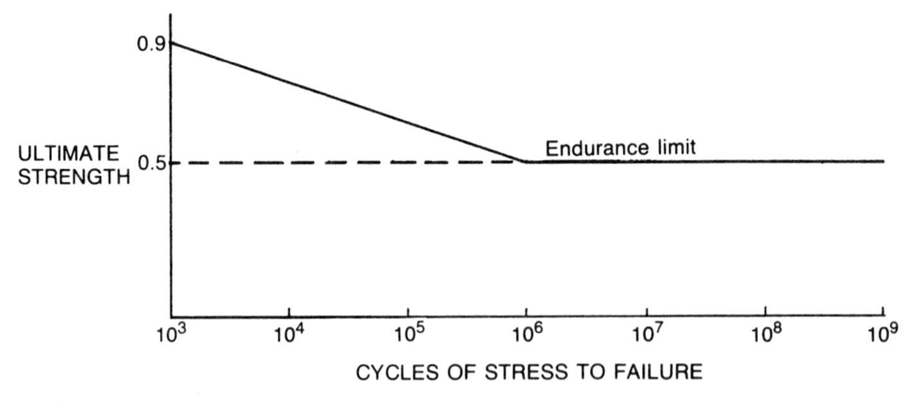

Figure 6.8 Generalised fatigue curve for steel

7 The Motor Shaft Extension

Before proceeding with the detailed design of the motor shaft extension, let us summarise the requirements in a design brief.

1 Design motor shaft extension suitable for operation at 250 V 50 Hz and 4-poles, i.e. 1500 rev/min synchronous = 1425 rev/min actual full-load speed. See comments in item 4 below.

2 Tentative dimensions of the shaft lengths are as in Figure 6.6, and the shaft assembly will be supported in two single-row ball bearings of about 20 mm bore.

3 Power take-off is likely to be via pulley, gear or sprocket, say 50 mm effective diameter, at the shaft end.

4 Because we should endeavour to standardise the size of the shaft extension for the entire range of variants, we will carry out our calculations on the model with the most onerous duty, i.e. that having the highest starting torque. This will be the variant with capacitor-start (developing 350% of full-load torque at starting), the highest power rating (370W), and the lowest speed (6-pole machine with a speed of 950 rev/min).

5 In the interests of economic manufacture, the shaft material will be free-cutting mild steel to British Standard Specification 220M07 (the former En1A), with the following properties

Youngs modulus of elasticity	200 $GN\,m^{-2}$
Ultimate tensile strength	386 $MN\,m^{-2}$
Ultimate shear strength	290 $MN\,m^{-2}$

6 Reserve factor,

6.1 Minimum value of relationship between yield stress and ultimate stress is 0.50 for 220M07 steel.

6.2 Factor of uncertainty would be for normal industrial material and processes, and would also allow for self-regulating thermal-overload cut-out to be built into the design to protect against repeated overload conditions, say 0.40.

6.3 Stress concentration.

If the shaft extension is to be plain cylindrical with no discontinuities, other than a well-radiused step-down in diameter from the bearing journal portion, we can estimate a stress concentration factor, following the procedure of Duggan (4).

If the shaft extension is to be supplied with a keyway, the stress concentration factor will be higher, and the reserve factor lower. Clearly, both possibilities must be examined.

6.4 Application of loading is sudden, and this can be accommodated in the calculations.

Plain shaft extension

Before proceeding, we must establish a number of probable relationships concerning the shaft extension geometry. To keep stress concentration at a minimum, we must also keep any shaft discontinuity at a minimum. The bearing journal diameter $d2$ must be slightly larger than the shaft extension diameter $d3$, so that when the interference fit ball bearing is being assembled, it does not score the shaft extension while being passed along it.

Refer to Figure 7.1. Let us set $d2/d3 = 1.05$, and the relationship $r/d3 = 0.2$ and this will produce a minimum step-down in diameter which is well radiused. Additionally, we need to consider some other aspects of the component before arriving at a reserve factor.

1 Type of loading.
Most fatigue data are derived from tests on specially prepared specimens, which are subjected to rotating bending loading, and an attempt is made to relate these data to the actual working conditions of a component by the application of a load correction factor $C1$.

2 Surface finish.
Test specimens are normally very accurately machined and are highly polished.

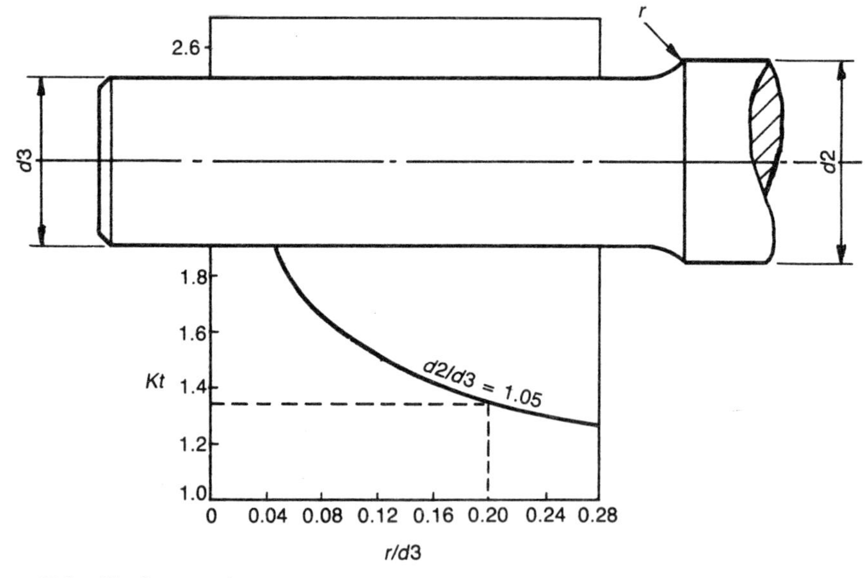

Figure 7.1 Shaft extension geometry

Component finishes, which may be sub-standard to this, can be allowed for by the application of a surface finish factor Ks.

3 Stress concentration.
Curves of stress concentration factors, related to $d2/d3$ and $r/d3$, are available in published literature for various forms of loading, e.g. bending, torsion, axial, and are designated Kt for direct stress and Kts for shear stress.

4 Component size.
Fatigue data obtained from accurate, highly polished specimens of 7.62 mm diameter, have to be adjusted to allow for variations in component size. Published factors are available, which are based on experimental work, and these are designated Cs.

Recent work carried out by NASA in the USA, on the design of shafts for longer life, also recognises the contribution of the reliability of endurance limit data, operating temperature, shaft duty cycle, i.e. start/stop cycles, transient overloads, vibration and shock loads, together with a number of other miscellaneous factors such as heat treatment, cold working, plating, welding, stress corrosion, fretting corrosion, etc.

The following nomenclature is used in determining the reserve factor for a plain cylindrical shaft extension, subjected to rotating bending.

Su = ultimate tensile strength of material = 386 MN m^{-2} for 220M07
Se = endurance limit in rotating bending at 10^6 cycles = 0.5 Su
Se' = corrected endurance limit at 10^6 cycles
Cs = material size factor = 0.85 (from published data)
Cl = load correction factor = 1 (rotating bending, so no multiple)
Ks = surface finish factor = 0.9 (from published data)
Kf = fatigue strength reduction factor = $1 + q(Kt - 1)Ks$
Kt = stress concentration factor = 1.34 (from published data)
$q = 1/(1 + a/r) = 0.882$
a = material constant depending on tensile strength = 0.4 (from published data)
r = radius of notch (shaft radius), say, 3 mm

$$Kf = 1 + 0.882(1.34 - 1)0.9 = 1.27$$

$$Se' = SeCsKsCl/Kf = 0.5 \times 386 \times 0.85 \times 0.9 \times 1/1.27 = 116.26 \text{ MN m}^{-2}$$

This is the maximum safe working stress to ensure no failure through fatigue and, as it is lower than the stresses determined by the relationship of yield to ultimate stress, or factor of uncertainty, it determines the overall reserve factor.

$$\text{Reserve factor} = 116.26/386 = 0.3$$

Keyways

We must now examine the effect of machining a keyway in the shaft extension. Again we have to decide some geometrical relationships in order to get started.

Using the relationships in Figure 7.2 enables us to establish a value for stress concentration in shear. The following nomenclature is used.

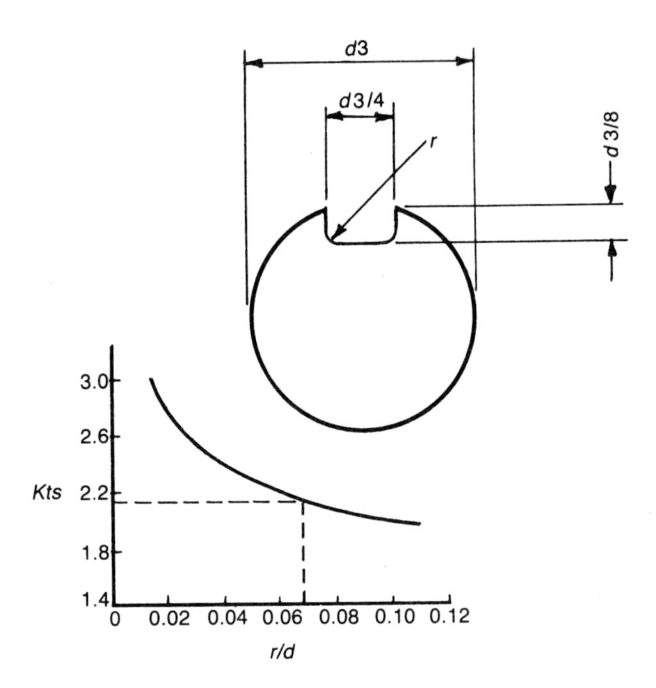

Figure 7.2 Keyway geometry

Ssu = ultimate shear strength of material = 290 MN m^{-2} for 220M07
Sse = endurance limit in rotating bending at 10^6 cycles = $0.5 Ssu$
Sse' = corrected endurance limit at 10^6 cycles
Cs = material size factor = 0.85 (from published data)
Cl = load correction factor = 0.58 (from published data)
Ks = surface finish factor = 0.9 (from published data)
Kfs = fatigue strength reduction factor = $1 + qs(Kts - 1)Ks$
Kys = stress concentration factor = 2.15 (from published data)
$qs = 1/(1 + 0.6a/r) = 0.806$
a = material constant depending on tensile strength = 0.4 (from published data)
r = radius of notch (keyway radius), say, 1 mm

$$Kfs = 1 + 0.806(2.15 - 1)0.9 = 1.834$$

$$Sse' = SseCsKsCl/Kfs = 0.5 \times 290 \times 0.85 \times 0.9 \times 0.58/1.834 = 35.08 \text{ MN m}^{-2}$$

This is the maximum safe working stress to ensure no failure through fatigue associated with the shaft extension keyway. As it is even lower than that for a plain shaft extension, it is the stress which determines the overall reserve factor, **for motors supplied with keyways in the shaft extension.**

So we must now calculate the sizes of the shaft extensions for both the plain cylindrical and for the keyway option.

Shaft extension diameter

For a plain cylindrical shaft extension, the lowest reserve factor we have calculated is 0.3.

$$\text{Safe working tensile stress} = 386 \times 0.3 = 116 \text{ MN m}^{-2}$$
$$\text{safe working shear stress} = 290 \times 0.3 = 87 \text{ MN m}^{-2}$$

$$\text{power} = 2\pi NT \text{ watts where } N = \text{shaft speed in rev/s}$$
$$T = \text{full-load torque in Nm}$$

$$\text{so that } T = (370 \times 60)/(2\pi \times 950) = 3.719 \text{ Nm}$$

Allowing for capacitor start and for sudden loading, i.e. start torque is 350% full-load torque and sudden loading doubles the applied force ...

$$\text{start torque } Ts = 3.719 \times 3.5 \times 2 = 26.03 \text{ Nm}$$

```
5 PRINT CLS
100 PRINT"MOTOR SHAFT EXTENSION
110 PRINT"_____________________
120 PRINT:PRINT
130 PRINT"THE DIAMETER OF A MOTOR SHAFT EXTENSION DEPENDS UPON MATERIAL,
140 PRINT"POWER, SPEED, TORQUE & STARTING MODE
150 PRINT
160 INPUT"YOUNG'S MODULUS (GN/M↑2) ";E
170 PRINT
180 INPUT"ULT TENSILE STRENGTH (MN/M↑2) ";B
190 PRINT
200 INPUT"ULT SHEAR STRENGTH (MN/M↑2) ";S
210 PRINT
220 INPUT"RESERVE FACTOR ";R
230 PRINT
240 INPUT"FULL LOAD POWER (WATTS) ";P
250 PRINT
260 INPUT"FULL LOAD SPEED (REV/MIN) ";N
270 PRINT
280 INPUT"PERCENT STARTING TORQUE ";A
290 PRINT
300 INPUT"SUDDEN APPLICATION? (Y/N) ";B$
310 IF B$="Y" THEN C=2
320 IF B$="N" THEN C=1
330 T1=P*60*A*C/(2*π*N*100)
340 D1=(16*T1/(S*R*π*1000000))↑(1/3)
350 PRINT"]"
360 INPUT"PULLEY (OR GEAR) DIAMETER (M) ";F
370 PRINT
380 INPUT"SHAFT EXTENSION LENGTH (M) ";L
390 T2=T1*2/F
400 M=T2*L
410 D=((((M↑2+T1↑2)↑(1/2)+M)*5.09/(1000000*B*R)))↑(1/3)
440 PRINT:PRINT
450 PRINT"SHAFT DIAMETER IS ";D*1000;"MM"
460 J=T2*L↑3/(600000000000*.049*D↑4)
470 PRINT
480 PRINT"SHAFT DEFLECTION AT START IS ";INT((1000000*J)+.5)/1000;"MM"
READY.
```

Figure 7.3 Program for shaft extension diameter and deflection

```
MOTOR SHAFT EXTENSION
_____________________

THE DIAMETER OF A MOTOR SHAFT EXTENSION DEPENDS UPON MATERIAL,
POWER, SPEED, TORQUE & STARTING MODE

YOUNG'S MODULUS (GN/M↑2)  200

ULT TENSILE STRENGTH (MN/M↑2)  386

ULT SHEAR STRENGTH (MN/M↑2)  290

RESERVE FACTOR  .3

FULL LOAD POWER (WATTS)  370

FULL LOAD SPEED (REV/MIN)  950

PERCENT STARTING TORQUE  350

SUDDEN APPLICATION (Y/N) Y

PULLEY (OR GEAR) DIAMETER (M)  .05

SHAFT EXTENSION LENGTH (M)  .08


SHAFT DIAMETER IS  19.57 MM

SHAFT DEFLECTION AT START IS  .124 MM
```

Figure 7.4 Printout for shaft extension (start-up conditions)

Force at gear effective diameter $F = 26.03/0.025 = 1041$ N
Shaft bending moment at drive bearing $M = 1041 \times 0.08 = 83.3$ Nm

As the shaft extension will be subjected to combined bending and torsion, we must design it so that the principal stress does not exceed the safe working stress already decided.

$$\text{principal stress} = \sigma/2 + \sqrt{(\sigma/2)^2 + (\sigma s)^2}$$
$$\text{where } \sigma = 32M/\pi d3^3$$
$$\text{and } \sigma s = 16Ts/\pi d3^3$$

This calculation indicates that we need a shaft extension of 19.57 mm diameter. Repeating the calculation for the steady-state full-load torque condition, shows that the shaft extension need be only 10.23 mm diameter. So we are faced with another decision. What size of shaft extension shall we specify?

Certainly, the extension diameter must not be less than 10.23 mm, for that is required to ensure we get no failures due to fatigue when running continuously at full-load. However, the start-up conditions, which give a torque of seven times the full-load value, last only for a fraction of a second. Furthermore they obtain only if the motor is under full-load or overload at starting. This high stress exists instantaneously and thereafter drops to that of normal load conditions. During the brief instant that the worst condition exists, there may be momentary overstressing of the shaft

```
MOTOR SHAFT EXTENSION

THE DIAMETER OF A MOTOR SHAFT EXTENSION DEPENDS UPON MATERIAL,
POWER, SPEED, TORQUE & STARTING MODE

YOUNG'S MODULUS (GN/M↑2)  200

ULT TENSILE STRENGTH (MN/M↑2)  386

ULT SHEAR STRENGTH (MN/M↑2)  290

RESERVE FACTOR  .3

FULL LOAD POWER (WATTS)  370

FULL LOAD SPEED (REV/MIN)  950

PERCENT STARTING TORQUE  100

SUDDEN APPLICATION (Y/N) N

PULLEY (OR GEAR) DIAMETER (M)  .05

SHAFT EXTENSION LENGTH (M)  .08

SHAFT DIAMETER IS  10.23 MM

SHAFT DEFLECTION AT START IS  .236 MM
```

Figure 7.5 Printout for shaft extension (steady-state running)

extension, but then the ductile material of the shaft will allow a redistribution of the stress, almost as it would with static loading. The high instantaneous stress is not repeated cyclically, nor is it reversed as in classic fatigue conditions, and so its effect is minimal. With this in mind we may ignore the overstress condition and treat the extension as for the steady running state.

Checks must be made on the shaft extension with a keyway. The sizes required for start and run conditions are 26.49 mm and 13.85 mm respectively.

Clearly, a 15 mm diameter shaft extension would be adequate for both plain and keyway variants under steady-state running conditions. BS5000 Part 11 allows for a shaft extension of 16 mm diameter as standard in the appropriate motor frame size (BM89). For those of nervous disposition, who may be worried about possible overstressing at start-up, this British Standard does allow for the specification of a 19 mm diameter shaft extension for certain applications. We will follow BS5000 and adopt a 16 mm diameter shaft extension for the current motor range.

Figure 7.3 shows a simple program for calculating shaft extension diameter and also maximum deflection occurring at the shaft end. Figures 7.4 and 7.5 show printouts for the 370 watt 6-pole machine on starting and during steady-state running.

8 The Motor Shaft

Having explored exhaustively the motor shaft extension and made the necessary decisions regarding its size, we must now turn our attention to that portion of the shaft, between the two ball bearings, where the rotor stack is positioned. Here a number of criteria must be satisfied.

Under any condition of loading, maximum deflection must not exceed 10% of the radial air-gap between rotor diameter and stator bore i.e. 0.03 mm. This limitation is necessary to avoid excessive unbalanced magnetic pull on the rotor.

Material stresses must be kept within the safe limit.

The critical speed (resonant frequency) of the shaft must be well outside the motor speed range of 0–1425 rev/min. (0–2850 rev/min. for 2-pole machines).

The rotor stack will be made up of a number of thin insulated discs, pierced around the periphery with holes to accommodate the rotor bars, and with a central hole to fit the shaft diameter we are about to calculate. At this point we are faced with a number of problems.

We have no idea of the number or size of the peripheral holes.

We know it may be necessary to provide ventilation holes through the laminations to assist cooling, but have no clue as to number and size.

We do not know what size the central hole for the shaft will be until the shaft diameter has been calculated.

We do not know the rotor stacking factor.

In spite of these imponderables, the designer has to determine the shaft diameter, so must make a number of reasonable assumptions. He or she will probably opt for worst case conditions.

Assume there are no holes in the lamination periphery; this means greater mass and hence more deflection.

Assume there will be no ventilation holes in the laminations; again more mass and greater deflection.

The central hole will eventually be filled up with the mtor shaft, so ignore it; again more mass and greater deflection.

Rotor stacking factor makes allowance for the impossibility of getting laminations so tightly compressed together that the resultant cylinder is of the same density as the constituent discs. It also allows for the lower density of the insulating material between the laminations. Assuming a stacking factor of unity gives the worst case condition here.

Assume the rotor stack to apply a point load to the shaft, rather than being distributed along it. This will also give worst case conditions for shaft deflection and stress.

Assume the force due to power transmission, acting on the shaft extension, is upward. Again, this gives worst case conditions for shaft deflection and stress.

Some of these assumptions may seem unnecessarily severe, but we must remember the designer is largely feeling the way in the dark, and these additional **reserve** factors may stand him or her in good stead. The final decison on shaft size can always be modified, to offset the worst case parameters, should the designer feel so inclined later in the design.

Free body diagram

We begin the design of the shaft with a free body diagram, which you will remember from your studies in dynamics of systems. For this part of the assignment we will stay with the mid-range variant we have been considering, i.e. 180W 4-poles.

The free body diagram in Figure 8.1 shows the shaft supported in the terminal end bearing R1 and the drive end bearing R2, carrying the rotor mass W at a distance $L1$ from the terminal end bearing. The power transmission force F, due to belt pull, chain pull, or gear thrust, is at distance $L3$ from the terminal end bearing. The distance between the two bearings is $L2$.

Dealing first with the rotor mass and the force which it exerts on the motor shaft

$$\text{force } W = \text{rotor volume} \times \text{material density} \times \text{acceleration due to gravity}$$

$$W = \pi/4(D2^2L) \times (0.0078) \times (9.81) = 24\text{N}$$

where material density = $0.0078 \text{ kg cm}^{-3}$

acceleration due to gravity = 9.81 m s^{-2}

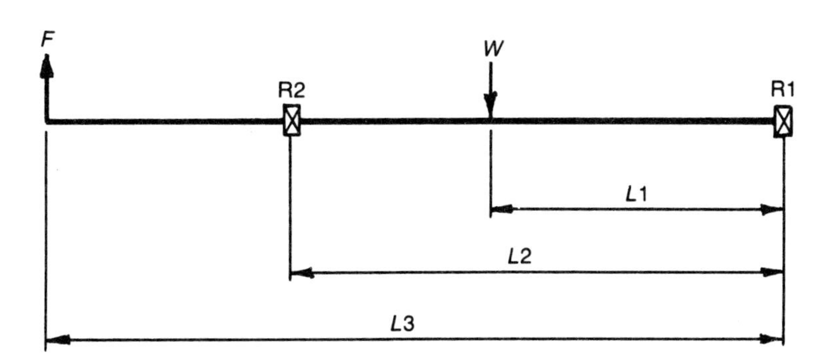

Figure 8.1 Free body diagram of shaft

Next we calculate the power transmission force on the shaft extension.

$$\text{power} = 2\pi NT \text{ from which } T = (180 \times 60)/(2\pi \times 1425) = 1.206 \text{ Nm}$$

Allowing for capacitor-start and sudden load application

$$\text{start torque } Ts = 1.206 \times 3.5 \times 2 = 8.443 \text{ Nm}$$
$$\text{at gear effective diameter, force } F = 8.443/0.025 = 337.74 \text{ N}$$

Shaft dimensions for the 180W 4-pole variant are

$$L1 = 70 + L/2 = 95 \text{ mm } (L = \text{rotor stack length of 50 mm})$$
$$L2 = 70 + L + 40 = 160 \text{ mm}$$
$$L3 = 160 + 80 = 240 \text{ mm (refer to Figures 6.6 and 6.7)}$$

Taking moments about R1

$$R2 = (W.L1 - F.L3)/L2 = -492 \text{ N (minus sign shows } R2 \text{ acts downward)}$$
$$R1 = R2 + W - F = 179 \text{ N}$$

MacAulay's method

To calculate shaft maximum deflection we use MacAulay's method.

Consider a plane at distance x from R1, then

$$\text{bending moment } EI\,\mathrm{d}^2y/\mathrm{d}x^2 = -R1.x + W(x - L1) + R2(x - L2)$$
$$\text{slope } EI\,\mathrm{d}y/\mathrm{d}x = C - R1.x^2/2 + W(x - L1)^2/2 + R2(x - L2)^2/2$$
$$\text{deflection } EIy = D + Cx - R1.x^3/6 + W(x - L1)^3/6 + R2(x - L2)^3/6$$

The constants of integration, C and D, are found by considering the points in the shaft where deflection is zero, i.e. at the two bearings R1 and R2, where x is 0 and 160 respectively.

$$\text{at } x = 0;\ EIy = 0 = D \text{ and } D = 0$$
$$\text{at } x = 160;\ EIy = 0 = 0 + 160C - 179(160)^3/6 + 24(65)^3/6$$
$$\text{from which } C = 756\,787$$

The position of maximum deflection occurs where the slope of the shaft is zero, i.e. where $\mathrm{d}y/\mathrm{d}x = 0$.

By inspection, this almost certainly occurs between R1 and W. However, let us first check to see whether it occurs between W and R2, i.e. is x between 95 mm and 160 mm from R1?

$$EI\,\mathrm{d}y/\mathrm{d}x = 0 = C - R1.x^2/2 + W(x - L1)^2/2$$
$$0 = 756\,787 - 89.5x^2 + 12(x^2 - 190x + 95^2)$$
$$0 = 756\,787 - 89.5x^2 + 12x^2 - 2280x + 108\,300$$
$$0 = 756\,787 - 77.5x^2 - 2280x + 108\,300$$
$$\text{from which } -77.5x^2 - 2280x + 865\,087 = 0$$

Applying the standard solution for a quadratic

$$x = (2280 \pm 16\,534)/-155$$

from which $x = 91.96$ or -121.38 both outside the range of 95 to 160

This confirms that maximum deflection is probably between R1 and W, i.e. x is between 0 and 95 mm from R1.

$$EI\,\mathrm{d}y/\mathrm{d}x = 0 = C - R1\,.\,x^2/2$$

$$0 = 756\,787 - 89.5x^2$$

from which $x = \sqrt{756\,787/89.5}$

and $x = 91.95$ mm from R1.

Substituting this value for x in the deflection formula, gives the value of maximum deflection in terms of E and I, the modulus of elasticity of the material and the second moment of area of the shaft cross-section. However, we know from our design brief that maximum shaft deflection must not exceed 0.03 mm and so we can determine the diameter of the shaft to suit this condition.

At $x = 91.95$

$$EIy = D + Cx - R1\,.\,x^3/6$$

$$EIy = 0 + 756\,787(91.95) - 179(777\,419)/6$$

$$EIy = 69\,586\,565 - 23\,193\,003$$

$$EIy = 46\,393\,562$$

for mild steel $E = 200\,000\ \mathrm{N\,mm^{-2}}$

for a cylindrical shaft $I = \pi d^4/64$

for this shaft $y = 0.03$ mm

shaft diameter $d = \sqrt[4]{46\,393\,562/(200\,000 \times 0.049 \times 0.03)}$

$d = 19.93$ mm

The shaft must be not less than 19.93 mm diameter in order that maximum deflection does not exceed 0.03 mm. The preferred size is 20 mm.

Stress

We must now check for material stresses induced in a shaft of 20 mm diameter by this form of loading, and the values are

actual bending stress $= 34.4\ \mathrm{MN\,m^{-2}}$
actual shear stress $= 5.4\ \mathrm{MN\,m^{-2}}$
actual principal stress $= 35.2\ \mathrm{MN\,m^{-2}}$

As will be seen, these stresses are extremely low, so that material stress is not a critical factor in the shaft design.

Critical speed

Next we must check shaft critical speed. When a shaft supported in bearings is rotated at increasing speed, it will reach a speed when it will become unstable and begin to whirl, rather like a child's skipping rope. This speed is known as the fundamental critical speed of the shaft. If shaft speed is held at this value, whirling will commence and the amplitude of the resulting vibrations will increase and may ultimately cause the shaft to fracture. If the speed is increased beyond the critical value, the shaft will resume stable rotation. The phenomenon of whirling occurs at every multiple of the fundamental critical speed and must be taken into account when designing rotating machinery.

A shaft supported in bearings is an elastic system and, as such, will vibrate at a resonant frequency. For any elastic system, the frequency of its natural fundamental vibration is given by

$f = (1/2\pi)\sqrt{g/y}$ Hz where g = acceleration due to gravity
y = static deflection in same units as g

Thus, if g is expressed as 9.81 $\mathrm{m\,s^{-2}}$ then

$f = (1/2\pi)\sqrt{9.81/y}$ Hz which is $(30/\pi)\sqrt{9.81/y}$ cycles/min.

and $f = 30\sqrt{1/y}$ cycles/min

So for a rotating shaft, the fundamental critical speed is

$f = 30\sqrt{1/y}$ rev/min where y is the maximum static deflection in metres.

Shaft size standardisation

As we are concerned with the design of a range of motors, we must perform these calculations for all twelve variants, and this is excessively time consuming. Once again we can turn to the microcomputer for help. Figure 8.2 shows a program for calculating motor shaft diameter and critical speed, and Figure 8.3 shows the printout for the 180W 4-pole variant.

Trade-off decisions

The results of the shaft diameter calculations are summarised in Table 8.1. This shows the shaft diameters required for the full range of motors according to machine speed

Table 8.1 Shaft diameter by variant

	120W	180W	250W	370W
2-poles	14.84	16.98	18.91	21.85*
4-poles	17.62	19.92	22.28	25.72
6-poles	19.43	21.97*	24.57	28.65

```
5 PRINT CLS
100 PRINT"MOTOR SHAFT SIZE
110 PRINT"─────────────────
120 PRINT
130 INPUT"ROTOR STACK DIA (MM) ";D2
140 PRINT
150 INPUT"ROTOR STACK LENGTH (MM) ";L4
160 PRINT
170 W=INT((9.81*π*(.1*D2)↑2*(.1*L4)*.0078/4)*100+.5)/100
180 INPUT"FULL LOAD POWER (WATTS) ";K
190 PRINT
200 INPUT"FULL LOAD SPEED (REV/MIN) ";S
210 PRINT
220 F=INT(((K*30/(π*S))*7/.025)*100+.5)/100
230 INPUT"LENGTH L1 ";L1
240 PRINT
250 INPUT"LENGTH L2 ";L2
260 PRINT
270 INPUT"LENGTH L3 ";L3
280 PRINT
290 R2=(F*L3-W*L1)/L2
300 R1=W+R2-F
310 C=(R1*L2↑3/6-W*(L2-L1)↑3/6)/L2
320 P=SQR(2*C/R1)
330 Y=P*C-(R1*(P↑3))/6
340 D=SQR(SQR(Y/(200000*.049*.03)))
350 PRINT"SHAFT DIA IS ";INT(D*100+.5)/100;"MM"
360 PRINT
370 R3=W*(L2-L1)/L2
380 C1=(R3*L2↑3/6-W*(L2-L1)↑3/6)/L2
390 P1=SQR(2*C1/R3)
400 Y1=(P1*C1-(R3*P1↑3)/6)/(200000*.049*(INT(D*100+.5)/100)↑4)
410 R=30*SQR(1/(Y1*.001))
420 PRINT"CRITICAL SPEED IS ";INT(R);" REV/MIN"
READY.
```

Figure 8.2 Program for shaft diameter and critical speed

```
MOTOR SHAFT SIZE

─────────────────

ROTOR STACK DIA (MM)
 89.36

ROTOR STCK LENGTH (MM)
 50

FULL LOAD POWER (WATTS)
 180

FULL LOAD SPEED (REV/MIN)
 1425

LENGTH L1?
 95

LENGTH L2?
 160

LENGTH L3?
 240

SHAFT DIA IS  19.92 MM

CRITICAL SPEED IS  26660  REV/MIN
```

Figure 8.3 Printout for shaft of 180W 4-pole variant

and power output. If we now *call back* some of the severe reserve factor, which we created by adopting worst case conditions for our shaft calculations, we can say that the shaft sizes range from 15 mm to 28 mm, both of which are preferred sizes. There are, however, some limiting features.

We have already settled on a common shaft extension diameter of 16 mm.

Ball bearings will be selected which have a bore diameter slightly larger than this 16 mm so as not to score the shaft extension during assembly; a 17 mm bore looks about right.

These bearings will need abutment shoulders on the shaft against which they will be located axially; ideally this should be 1.5 mm radial depth, giving a minimum shaft diameter of 20 mm.

All shafts contained within the **box** in Table 8.1 will have to be made from a shaft size of 20 mm for the reasons mentioned above; over the five year period of manufacture, this will be around 108 700 shafts.

The remaining 96 300 shafts will need to be of five different sizes,
22 mm for the two items marked * in Table 8.1
different diameters for the remaining four variants

Clearly, six different shaft sizes could lead to enormous problems of logistics.

Six different stocks of raw materials and six different stocks of finished components will inevitably lead to mix-ups, to shortages and to lost production.

Six different stocks of rotor laminations would also be required because of the differences in centre hole size, together with six different sets of lamination press tools; all this is highly expensive and subject to mix-ups and shortages.

Only two possibilities are tenable. Either we have one standard shaft diameter at 28 mm or we have two, one at 20 mm for those variants in the box and one at 28 mm for the rest. The single shaft size at 28 mm has these advantages ...

no extra tooling or maintenance costs
no extra logistical problems

and the disadvantage of ...

giving away redundant material on all shafts inside the box, i.e. on 108 700 units over the five year period

One shaft size at 28 mm and another at 20 mm has the advantage of ...

no give-away material on 108 700 shafts

and the disadvantages of ...

extra expense for tooling and maintenance
extra logistical problems

If we carry out a rough costing on these two alternatives, we might get something like this ...

for one shaft size of 28 mm	£
extra tooling and maintenance costs	nil
extra logistics expense	nil
give-away material on 108 700 shafts	28 000
extra machining cost on 108 700 shafts	21 000
total extra cost of this alternative	49 000

for two shaft sizes of 28 mm and 20 mm	
extra tooling and maintenance costs	20 000
extra logistics expense estimated at	15 000
give-away material on 108 700 shafts	nil
extra machining cost on 108 700 shafts	nil
total extra cost of this alternative	35 000

On the surface, the decision ought to be in favour of the second alternative, two shafts. However, there are two further considerations.

The expense of logistics foul-ups is a rough estimate and could be well adrift of the real costs.
The difference between the two alternatives, as costed above, works out at about 7 pence on every shaft over the five years.

On further consideration, it is clear the designer would opt for a slightly more expensive motor shaft in order to ensure freedom from extra logistics problems over the five year period.

This has been an important intermediate decision, which sprang from the fundamental decision to manufacture a **range** of variants. The designer has decided on a trade-off between product cost and manufacturing convenience. By opting for a slightly more expensive motor shaft, an extra 7p to be absorbed from profit, the designer hopes to avoid ...

split-ordering of shaft raw materials
extra material cost due to smaller quantities required in each size
segregated stocks of raw materials
separate tooling requirements
dual manufacturing runs
extra component expense due to smaller batch quantities of each size
segregated component stocks
component shortages due to changed schedule requirements

All the above generate extra costs. With two shaft sizes, logistics problems would be present throughout the life of the product. With only one shaft size, manufacturing is made easier throughout the lifespan of the product, at a very small drop in profit margin. It appears this decision is well justified.

Bearing selection

In selecting a suitable bearing for the motor shaft, the basic considerations are...

bore to be slightly larger than shaft extension to avoid scoring during assembly; as already indicated a bore of 17 mm will do

bore to be 3 mm smaller than shaft diameter to allow for bearing abutment shoulder

bearing should be as small as possible so as not to encroach on the area around the end frame boss, which is required for cooling air intake

the smaller the bearing the cheaper it is

the bearing must satisfy the load rating required under dynamic conditions for acceptable bearing performance and life

to avoid logistics problems, a bearing should be selected which can be common to both ends of the shaft.

The primary function of a bearing is to support a load while rotating for as long as the equipment, of which it is a part, continues to function. The accurate determination of forces acting on a bearing is often quite impossible, so approximate calculations and generous reserve factors are usual. In the case of ball bearings, the geometry of the bearing itself may affect the way in which the onset of failure is resisted. Failure is usually characterised by the visible signs of fatigue, such as spalling and flaking of the races and the balls. Once these signs are evident, the bearing should be replaced to avoid complete collapse.

It is not possible to reliably predict the exact life of any bearing because of the random nature of a number of contributory causes of failure. The homogeneity of the material is subject to variability; balls, races and cages cannot be made to identical sizes and tolerances; heat treatment will vary from batch to batch; finishes will not always be similar; etc., etc.

The scatter of bearing lives has given rise to the internationally agreed 'B-10 life expectancy' of a bearing, which is defined thus...

> the B-10 life of a bearing is that life expectancy during which 90% or more of a given group of bearings, under specific loading conditions, will still be in service.

There are two attributes which are important in assessing bearing performance; they are...

basic dynamic capacity usually designated *C*, and
basic static capacity usually designated *Co*.

They are defined thus...

> The basic dynamic capacity of a bearing is the load that it will sustain for three million revolutions while operating at 33.3 rev/min.

Table 8.2 Bearing characteristics

Bearing type	Bore	Outer diameter	Width	*C*	*Co*	Limiting speed
light	17	40	12	7380N	4360N	18000 rev/min.
extra light	17	35	10	5250N	3020N	19000 rev/min.

The basic static capacity of a bearing is the load that may be applied to produce a deformation of 0.0001 of ball diameter of the most highly stressed ball.

Bearing life is approximately equal to $(C/R)^3$ where R is the applied journal load.

Static and dynamic capacities of bearings are quoted by bearing manufacturers in their catalogues. A typical example is shown in Table 8.2 for two bearings with 17 mm bores, one in the light series and the other in the extra light series.

The outer diameter of the bearing will be fairly critical as it affects the amount of free space around the end frame central boss which will be available for cooling air intake. The designer will certainly choose the extra light series bearing. It is 5 mm smaller in its critical outer diameter, its static and dynamic capacities are comfortably adequate, and its limiting speed is well above the operating range of the fastest variant, the 2-pole machine at 2850 rev/min.

We will standardise on this bearing for use at each end of the motor shaft, and this will avoid logistics problems in the component stores, both in the manufacturing plant and also in its service agencies.

9 Motor Casing

The next fundamental decision concerns the choice of material for the motor casing. In making this choice, the designer is determining which manufacturing resources will be used and is also committing their use in advance.

Functions

The functions to be performed by the motor casing are...

hold the stator stack in correct axial and longitudinal alignment
provide accurate locations for the two end frames
dissipate heat from the stator/winding assembly
support a nameplate with details of motor identity and classification
provide an anchor for the motor foot
meet safety requirements vis-a-vis operator and motor
prevent ingress of falling water into the motor.

The first two of these functions require accuracy and rigidity. The third function is achieved by both convection from the casing surface and by the passage of cooling air through the motor structure. The fourth function requires the provision of a stable base; the fifth also requires a stable base and in addition it must be capable of resisting the force due to starting torque. The sixth must ensure electrical and mechanical safety; for the former the casing must be insulated from electrical supplies and be earthed against accidental breakdown, for the latter any apertures in the casing must not permit entry and contact with any internal parts by the 'international finger'. This is merely an internationally agreed gauge which is designed to eliminate any possible contact between a human finger and any internal feature presenting either electrical or mechanical hazard. The final function must offer protection to internal features from water falling on the motor casing.

Almost any material could be considered a candidate for the casing, particularly some of the modern composites.

Alternative materials

The two traditional materials are die-cast aluminium and welded steel sheet. It is of these that we will examine the pros and cons.

Die-cast aluminium advantages
intimate contact with stator stack for good heat transfer

high thermal conductivity 0.57
light weight, density 2560 kg m^{-3}
very easy to machine
no separate fixing to stator stack required, cast in-situ
could be self-finish thus avoiding expense
can be made any shape in cross-section*

Die-cast aluminium disadvantages
high tooling costs add to special-to-type expense
high material cost, 6.8 times that of steel
greater material volume required for item strength
soft and easily damaged
process subject to porosity and surface blemishes

Welded steel advantages
low tooling cost gives low special-to-type expenses
low material cost, 0.15 times that of aluminium
lower material volume required for item strength
robust and not easily damaged
material is homogeneous

Welded steel disadvantages
relatively poor thermal contact with stator stack
low thermal conductivity 0.12
higher weight, density 7800 kg m^{-3}
can only easily be made with cylindrical cross-section*
less easy to machine
must be separately secured to stator stack
needs anti-rust finish

*A word about the two items above concerning casing cross-sectional shape. With a cylindrical shell, any motor tiebolts connecting the two end frames must pass through the stator stack thereby modifying the magnetic field in the region of the tiebolts. This automatically reduces motor performance, which can then only be restored by increasing the outer diameter of the stator stack. This adds considerable expense, e.g. wider lamination strip, larger press tools, larger motor frame, etc. With a contoured shell, say roughly square in cross-section, the motor tiebolts may pass outside the stator stack, through the corner tunnels, without causing any material redundancy. The corner tunnels also mightily improve the flow of cooling air through the motor.

Material selection

Choosing the appropriate casing material can perhaps be helped if some of the above pros and cons can be quantified. Consider the attributes required in the finished product, in their priority order, and try to set ratios for the various factors wherever possible, even though some may be only guesses.

Performance	steel	aluminium
thermal conductivity	1	4.75
ease of machining, say	1	1.25
physical strength and homogeneity	3	1
effective air ventilation	1	2
overall ratio 4 to 1 in favour of aluminium	3	11.88

Important! It should be noted that the overall ratio is based on the **product** of the various factors and not on their sum. The total factor for steel is $1 \times 1 \times 3 \times 1 = 3$. The total factor for aluminium is $4.75 \times 1.25 \times 1 \times 2 = 11.88$. We are again adopting the pairs-comparison technique, using our best judgement in order to decide the superior of the two materials and the ratio between them. This is totally subjective, but is the best we can do at this time.

Reliability	steel	aluminium
resistance to failure and homogeneity	2	1
structural rigidity	1.5	1
resistance to wear of spigot diameters	1.5	1
overall ratio 4.5 to 1 in favour of steel	4.5	1

Safety		
resistance to bursting and penetration, say	1.1	1
overall ratio 1.1 to 1 in favour of steel	1.1	1

Cost		
raw material	1	6.8
mass	3	1
extra material volume required, say	1	1.5
tooling (extra s-t-t expense)	1	1.2
extra operation, fixing stator in casing	1.2	1
overall ratio 3.4 to 1 in favour of steel	3.6	12.1

Note! In the case of cost, we are looking for the overall factor which is lowest, i.e. where cost is a minimum. So in this analysis, the extra cost incurred in using aluminium would be about 3.4 that of using steel, and clearly steel is the preferred alternative.

Maintainability
overall ratio 1 to 1, both equally maintainable

Life	steel	aluminium
longevity of casing, say	1.2	1
overall ratio 1.2 to 1 in favour of steel	1.2	1

Interchangeability
overall ratio 1 to 1, both equally interchangeable

Appearance		
general shape and texture possibilities	1	2
match between casing and end frames	1.5	1
overall ratio 1.3 to 1 in favour of aluminium	1.5	2

Some of these factors are factual, e.g. material costs, material mass, etc. Others are our best guesses and are subject to estimating inaccuracy. However, having looked again at our comparisons above and decided they are reasonable, we can use them to decide the values of the primary decisions in the forced-decision analysis program.

On the face of things, this form of decision making may seem unduly mechanistic and numerate, but consider how you would have made the decision between materials using intuitive methods. Intuitive processes would be almost identical to those in use here. We would consider the relative merits of each of the materials vis-a-vis the attributes we have to satisfy, i.e. which material has the greater strength, resistance to degradation, best safety features, most attractive appearance possibilities, lowest cost, etc., etc. Without consciously assigning numerical values to each we would, nevertheless, select one of the two for its overall superior qualities by subconscious comparison. With the present method, we merely bring these comparisons to the surface, examine them critically, then give them numerate values. What is more, these values are committed to paper and remain available for future examination and debate if required.

The intuitive method may be acceptable for only two alternative materials, but it would be quite inadequate if the choice had to be made between half a dozen materials. The rigorous discipline of the decision-forcing technique produces a more professional approach.

Running suitable values through the decision-forcing program produces a decision marginally in favour of welded steel, by around 1.2 to 1. Of course, the designer may still modify this decision should he or she choose; the designer is still the person with the ultimate responsibility for making the design decisions.

One immediate result of this fundamental decision on casing material, is an intermediate decision affecting the flow of cooling air through the motor. With a light alloy casing, the airflow would be in at the drive end frame, then through the corner tunnels of the casing and out through the terminal end frame.

With a cylindrical steel casing, however, the stator stack acts as a barrier to airflow between the two end frames. So air must be drawn in through the drive end frame and expelled from the same end of the motor assembly. Also air must be drawn in through the drive end frame and expelled from that end of the motor assembly. Provision must be made for air to pass freely through both end frames but, because of the necessity to prevent the ingress of falling water, the ventilation apertures must be suitably drip-proofed. To improve the airflow, it is permissible to pierce apertures in the 'overhangs' of the steel shell, provided such openings are below the horizontal centreline, to ensure non-ingression of falling water. They must also be of such size and shape as to prevent the insertion of the international finger. Having decided the material and probable shape of the casing, we should now move on and look at the design of the two end frames.

10 Motor End Frames

The two motor end frames pose almost the most stringent restrictions on the designer. They have to be strong enough to perform the power transmission function, open enough for free passage of cooling air, and packed with control equipment at the terminal end. Let us examine the design brief in detail.

Functions

Between them the two motor end frames perform a number of important functions. They must ...

enclose the internal features of the motor
locate shaft in its bearings coaxially with the stator bore
allow lubrication of the bearings
prevent ingress to the motor of falling water
allow circulation of cooling air
assist dissipation of heat from the motor
resist deformation and deflection by the operational forces
act as a safety barrier
enclose terminal board and connections with access as required
accommodate tiebolts for motor assembly
be rotatable to allow floor, wall and ceiling mounting
support static portion of start-winding switch

The casting process

Traditionally, end frames are castings either of iron or light alloy, iron being used where high operating forces are involved. In the present case light alloy will be quite adequate and, although more expensive than iron, it has superior thermal conductivity and will help achieve lower running temperatures. Before proceeding with design, it might help to review the casting process, so that the designer is aware of areas needing special attention. It is absolutely vital that the designer understands the processes and methods of manufacture. The process of casting consists of five phases ...

an enclosed cavity is filled with liquid component material at an elevated temperature, around 925K for non-ferrous metals

the air in the cavity is allowed to escape as the component material flows in

when the cavity has been filled, the component material is allowed to cool so that it sets and takes up permanently the shape of the cavity as the component material cools, it shrinks and occupies less volume than when hot

when set, the component must be removed from the cavity without damage to either cavity or component.

When designing any casting there are several points that must be conisdered in the interests of quality components...

keep component geometry simple
avoid sudden changes in component section thickness
allow surplus material for machining where precision is required
allow generous fillet radii
avoid large unrelieved surfaces
provide sufficient draft (taper) for component removal from die
keep component tolerances wide, subject to fit, form and function

These points and others can best be illustrated by looking at the design of a rudimentary end frame. Figure 10.1 shows the basic requirements of a motor end frame. It is a roughly bell-shaped object with a large central boss to accommodate the shaft bearing. As drawn this component has several very bad features.

a large unrelieved flat surface in the disc portion
a heavy cylindrical section in the central boss
sudden changes of section thickness
sharp internal and external corners
no draft angles

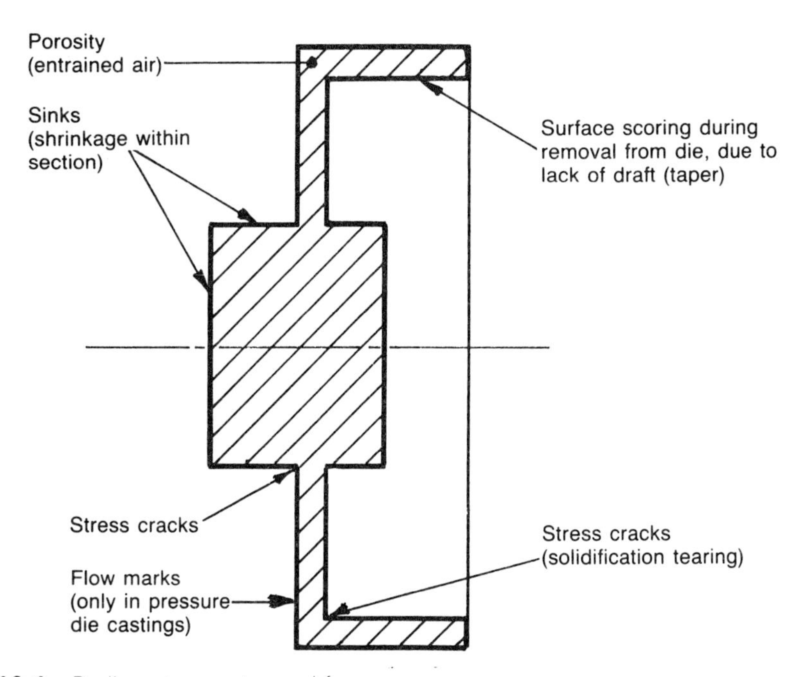

Figure 10.1 Rudimentary motor end frame

These bad features will cause weaknesses in the manufactured component.

sinks on the faces and diameter of the central boss
flow marks on the faces of the disc
porosity in the boss and at the outer diameter of the disc
stress cracks at the outer and inner diameters of the disc.

Sinks

Sinks are depressions in the surface of the component which are formed when the hot metal cools. Cooling takes place unevenly, due to some parts of the die cooling faster than others, and the last portions of the component to set act as reservoirs of molten material to feed those portions which set first. The mechanism of sink formation is illustrated in Figure 10.2. As the liquid metal begins to solidify, the thin section becomes plastic while the thick section is still largely fluid. Liquid material is drawn from the thick into the thin section to make up the reduction in volume due to shrinkage. As cooling continues, the fluid material in the thick section continues to contract. But its outer surface has by now become plastic and cannot flow, so internal stresses are created in the component which result in local collapse or sink of the surface.

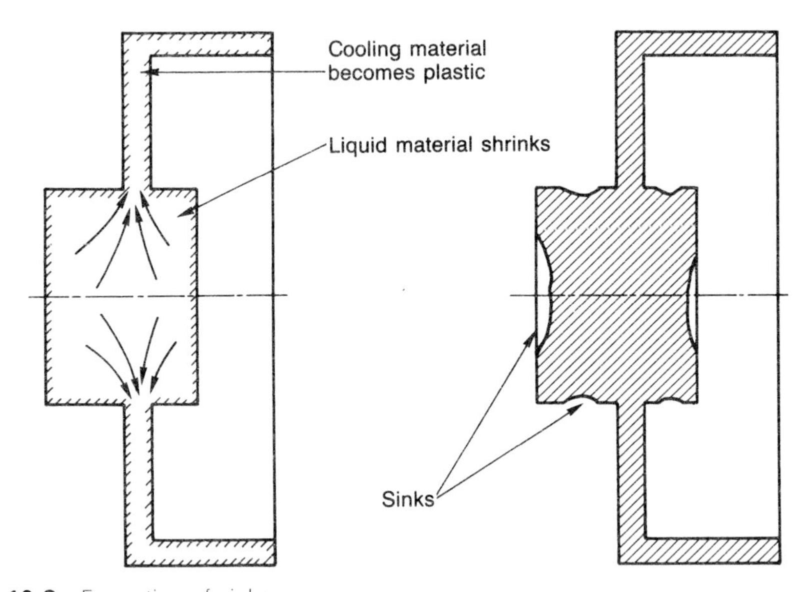

Figure 10.2 Formation of sinks

Porosity

Porosity is usually caused by turbulent flow in the hot metal. Whenever the hot material flowing through the die comes up against a sharp corner or a sudden change of section, the flow characteristic is changed. Turbulence is created and some of the air in the die becomes entrained (mixed) in the hot material. This air is unable to escape from the more dense component material and when the component sets, it will contain porous areas which have the consistency of a sponge. Such areas are sources of great weakness in the finished component. Porosity can also be caused by introducing the component material into the cavity in the wrong place. If the hot material enters at the top of the cavity, it will fall freely causing turbulence and entraining air. The more designers are aware of the limitations of the manufacturing processes, the better able they are to design components for trouble-free toolmaking.

Draft angles

Draft, or taper, on the cavity walls is required to ease the removal of the set component from the die. The finished component is always removed from the cavity while it is still quite hot, and any excessive force used may cause distortion in the component and damage to the cavity surface. Draft angles are essential if the component is to be removed without surface damage. It is important to remember that the component contracts on cooling, that is all features move toward the physical centre of the component. Therefore, it will tend to pull away from the **female** parts of the cavity, while gripping ever more firmly all **male** features such as pins, cores and inserts. For this reason it is usual to allow greater draft angles on male features than on female ones. The actual amount of draft will depend upon the shape and size of the component, and also on the component material. The material is pertinent, as this determines the temperature at which the casting operation is carried out, and thus how

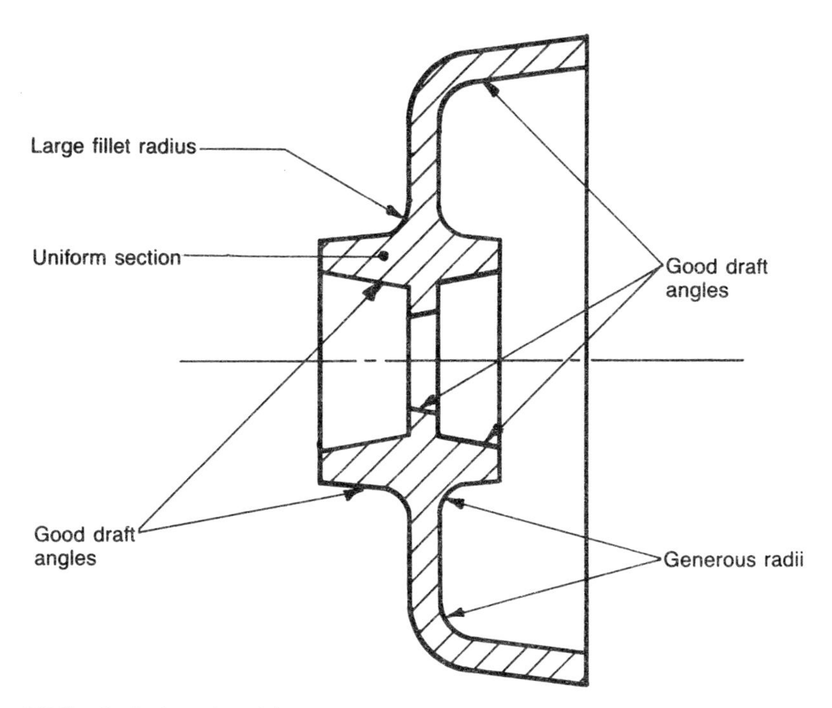

Figure 10.3 Redesigned end frame

much shrinkage is likely to occur. Most engineering textbooks give specific draft angle recommendations for various casting materials.

Figure 10.3 shows the end frame redesigned to incorporate the features discussed above; in this the draft angles have been exaggerated for clarity.

Die design

A word or two here about die design would be appropriate. It is obvious that the satisfactory design of cast components requires the designer to understand the casting process fully, and good die practice is a pre-requisite of good casting design. What follows applies principally to pressure die casting, where the hot component material is introduced into the die under considerable pressure. It is also true in some degree for gravity die casting, where the hot component material is poured by hand into the die at atmospheric pressure, and the weight of the molten material ensures cavity filling.

Dies should be structured to assist the smooth, turbulence-free flow of the hot component material. In many cases this means introducing the molten metal into the lower part of the cavity, probably near the geometric centre, and allowing it to flow upwards and outwards pushing the cavity air before it. It also means allowing sufficient vents in the die to enable this air to escape without a significant build-up of pressure within the cavity. At the same time, the vents must be of such size and shape as not to allow the molten material to get out of the cavity. It may also mean introducing reservoirs of hot material, known as **risers**, strategically placed within the die. These help prevent the formation of sinks. Also they help reduce the amount of shrinkage by keeping a constant supply of molten material readily available during cooling.

Single and multi-cavity dies

For precision castings, a single-cavity die will ensure maximum repeatability of component shape and size, providing casting temperature is controlled within accepted limits. Where large numbers of the same component are required, it may be necessary to sacrifice some precision for the better economics of multi-cavity dies. Here, the operation of, say, a six-cavity die will produce in one-shot, six times the output of a single-cavity die. Clearly, the multi-cavity die will require the use of a larger, and therefore more expensive, die casting machine, so the overall costs are influenced by factors other than just quantities required.

Because multiple cavities will vary in shape and size, one from another, due to variability in the toolmaking processes, there will be corresponding differences in the components produced from multi-cavity dies. However, there are rules which, if properly observed, will help to minimise this component variability.

Cavities should be precision machined to match each other as closely as possible for geometry. This has only to be done once, during die manufacture, and it is worth spending time and money on this to ensure that, perhaps, hundreds of thousands of components will be almost identical and require minimal corrective machining. With modern CNC (computer numerically controlled) die-sinking machines, the control of cavity shape and size is much closer than with the former manually controlled equipment.

Correct entry of molten component material into the cavity through a properly designed access, known as a **gate**, is essential. Eccentric gates tend to induce quite high internal stresses in the solidifying material and may cause subsequent component distortion.

Molten material is conveyed within the die, from the material feed point or points to the individual cavities, via channels known as **runners** in the die blocks. Runner layout is important to ensure that each cavity is filled simultaneously, thus guaranteeing identical casting conditions. With modern CAD (computer aided design) terminals, software is available which visually analyses dynamic material flow in dies and allows the designer to modify runner layout and cross-section for optimum material flow to all cavities.

Temperature control of both component material and die blocks is vital to avoid variability in the product. Dies may be either air- or water-cooled, and the aim should be to have a generally even temperature throughout the die blocks. Accurate control of temperature also helps to avoid **cold shuts**. These are caused by the incomplete uniting of separate material flows when they meet up in the die cavity. Clearly, cold shuts are visual blemishes as well as sources of weakness in the cast component.

Always design the die so that it is readily adjustable to compensate for burning and wear during service. All male features within the cavities should be manufactured toward the top limit of tolerance, so that wear and re-machining during life does not bring the feature below the bottom limit of size. Similarly, all female features should be manufactured toward bottom limit, to allow for in-life cleaning up. It will be obvious that the wider the tolerances on the cast component, the longer will be the life of the die between major overhauls. Wherever possible, features which are likely to be subject to frequent re-machining, should be made removable (as inserts) in order to preserve the main structure of the die for as long as possible. In turn, this means more components from the die and a lower tooling charge to be amortised over the component cost.

Selection of the correct component material is important. Alloying elements in aluminium casting materials can assist hot metal flow, thus avoiding die abrasion and ensuring much more faithful reproduction of cavity shape by the better filling of small features. Surface blemishes such as flow marks are also minimised by better flow characteristics.

From the foregoing it can be seen that change in the direction of flow of the hot material is a major cause of problems. So it might lead to sounder end frame design if the component was reduced to a plain disc with a central boss. This design strategy would also give the end frame greater resistance to deflection and promote more accurate location for the motor assembly vis-a-vis the stator bore. However, such a design would mean longer overhangs at each end of the motor casing, to make up for the missing end frame flanges, with a consequent increase in the difficulty and expense of coil winding. Nothing is for nothing, and another trade-off looms!

11 Stator and Rotor Laminations

We are now moving into the heart of the motor design. To give the motor life, we must add to the mechanical framework some electrical muscle. *Do not panic.* Most of what follows is also mechanical in essence, being concerned with establishing numbers and sizes. As the reader will by now have gathered, much of the work involved in small-power motor design is based on empirical data derived from extensive trial-and-error activity over many years. Even the pundits of motor design cannot always explain the theoretical lineage of their calculations. They know only that the results work in practice, and that, in the final analysis, what really matters is satisfying the needs of the customer, and turning manufacture into profitable income. The heart of the motor consists of two sets of laminations and two sets of windings. Let us begin by looking at the laminations.

Stator laminations

The overall sizes of the stator have already been settled; 152 mm outer diameter, 90 mm bore, 50 mm length for the mid-range variant. The details of stator teeth and slots must now be determined, and this is a purely mechanical activity.

The stator of an induction motor does not have salient (fixed) poles. It is designed so that several different numbers of poles can be accommodated in a single, common lamination. Figure 11.1 shows a typical stator lamination together with its rotor lamination. Both these items are normally punched from the same piece of strip material in a progression press tool.

The layout of a typical lamination tooth and slot arrangement is shown in Figure 11.2 (Data sheet 19 sheet 3) and this should be referred to as the design proceeds.

There are several factors to be considered when deciding the number of slots in a stator lamination. Some of these factors are enumerated here but for the timid, they may be largely ignored. They are included here to underline reasons for some of the decisions to be made.

Number of stator slots

A large number of stator slots reduces leakage reactance by reducing slot and zig-zag leakages. This mean more output from a given stator size, more breakdown torque, and somewhat better power factor and efficiency. A large number of stator slots also reduces troubles deriving from field harmonics, and tends to lower the intensity of magnetic noise while, at the same time, increasing the pitch of the noise components.

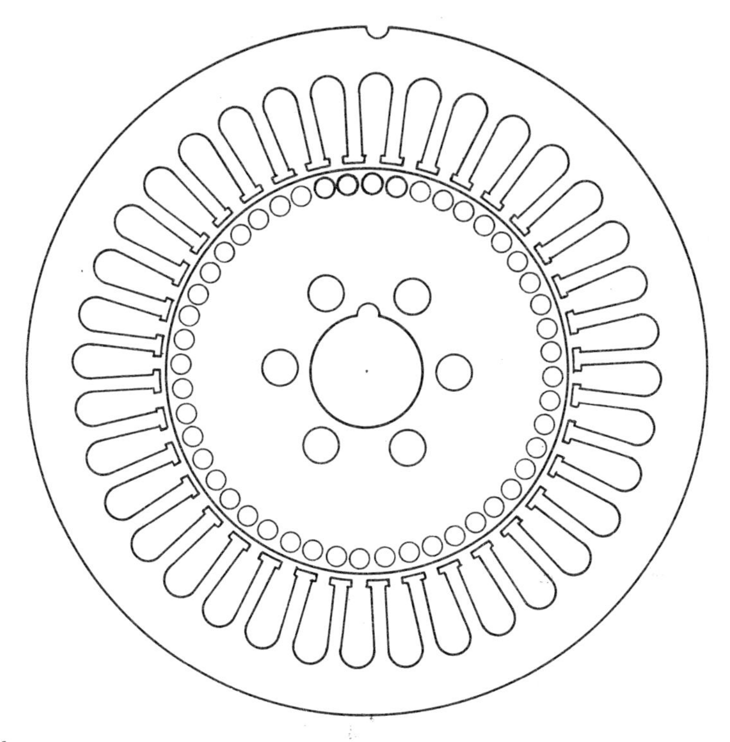

Figure 11.1

However, as the number of stator slots is increased, the space factor (room for windings) becomes poorer. There is always an upper limit to the number of stator slots, which is dictated by the physical size of the lamination.

A small number of stator slots reduces the cost of winding the stator coils and, generally, gives a better slot space factor. The minimum number of slots is usually governed by performance requirements, for too few would result in very wide but shallow slots carrying an excessive amount of copper wire.

Table 11.1 gives the recommended values for the number of stator slots (S) for various numbers of poles (P) when commencing stator design.

Figure 11.3 (Data sheet 19 sheet 4) shows the sequential procedure for stator lamination design, and some comments are appropriate.

Table 11.1 Number of stator slots

P	Recommended values of S											
2	6	12	18	24	30	36	42	48	54	60	66	72
4		12		24		36		48		60		72
6			18			36			54			72
8				24				48				72

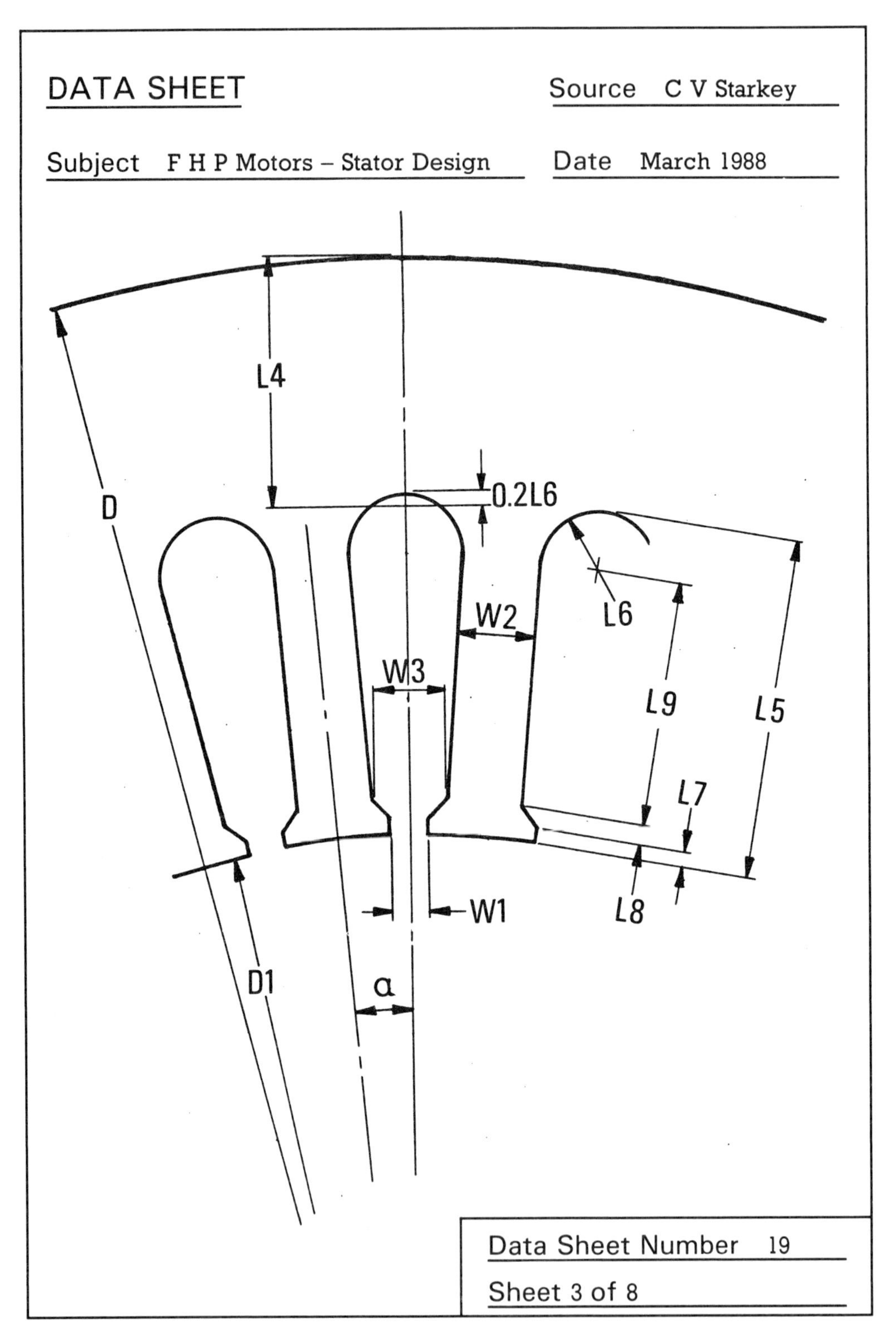

Figure 11.2 Data sheet for stator design

DATA SHEET

Source C V Starkey

Subject F H P Motors – Stator Design

Date March 1988

Sequential procedure for designing the stator lamination:-

1. D = Stator outside diameter – refer to sheet 1
2. D1 = Stator bore – refer to sheet 1
3. S = Number of stator slots – refer to sheet 2
4. W1 = Slot opening – refer to sheet 2
   if $S = 24$: $W1 = 0.68 + 0.0175D1$ mm
   if $S = 36$: $W1 = 0.38 + 0.0175D1$ mm
   if $S = 48$: $W1 = 0.0175D1$ mm
5. L7 = Tip depth – usually made 0.75 mm
6. L8 = Throat depth – usually made 1.5L7 – refer to sheet 2
7. W2 = Tooth width – refer to sheet 2
   for 2 poles: $W2 = (1.10 + 0.003D1)D1/S$ mm
   for 4+ poles: $W2 = (1.27 + 0.0035D1)D1/S$ mm
8. L4 = Yoke depth – refer to sheet 2
   $L4 = 1.15(S\ W2)/\pi P$ mm
9. α = Half slot angle $= 360^\circ/2S$
10. Constant $A = 0.5(D1 \sin\alpha - W2)$ mm
11. Constant $B = 0.5(D - D1) - L4$ mm
12. L5 = Slot depth $= \dfrac{B(1 + \sin\alpha) + 0.2A}{1 + 0.8 \sin\alpha}$ mm
13. L6 = Slot radius $= \dfrac{A + B \sin\alpha}{1 + 0.8 \sin\alpha}$ mm
14. L9 = Throat to radius $= L5 - (L6 + L7 + L8)$ mm
15. W3 = Minimum slot width $= 2(L6 - L9 \sin\alpha)/\cos\alpha$ mm

Data Sheet Number 19

Sheet 4 of 8

Figure 11.3 Data sheet for stator design calculations

slot opening W1 — equations were derived empirically from practice; the opening should be wide enough to permit easy insertion of the largest winding wire envisaged, after making allowances for slot stagger, insulation, and wire guides (if the stator coil is machine wound).

throat depth L8 — should be small enough to keep down leakage reactance and give maximum slot area, but large enough to give adequate tooth strength and to prevent magnetic saturation of the tooth tips.

tooth width W2 — these values give satisfactory starting points for design; it may be necessary to modify this dimension either way, during the later calculations for the winding details.

yoke depth L4 — this equation gives the yoke depth necessary to preserve a given ratio of tooth to yoke flux densities: the yoke depth is taken as actual iron

```
5 PRINT CLS
100 PRINT"STATOR LAMINATION DESIGN
110 PRINT"————————————————————————
120 PRINT:PRINT"USE WITH DATA SHEET 19 PAGE 3
130 PRINT
140 INPUT"NUMBER OF POLES ";P:PRINT
200 INPUT"STATOR OUTER DIAMETER (MM) ";D:PRINT
210 INPUT"STATOR BORE (MM) ";D1:PRINT
220 INPUT"NUMBER OF STATOR SLOTS ";S:PRINT
230 IFS=24THEN300
240 IFS=36THEN310
250 IFS=48THEN320
260 GOTO220
300 W1=.68+.0175*D1:GOTO350
310 W1=.38+.0175*D1:GOTO350
320 W1=.0175*D1:GOTO350
350 PRINT"SLOT OPENING W1 = ";INT(W1*100+.5)/100;"MM":PRINT
360 L7=.75
370 L8=1.5*L7
380 PRINT"THROAT DEPTH L8 = ";INT(L8*100+.5)/100;"MM":PRINT
390 IFP=2THEN420
400 IFP=4THEN430
410 IFP=6THEN430
420 W2=(1.1+.003*D1)*D1/S:GOTO450
430 W2=(1.27+.0035*D1)*D1/S:GOTO450
450 PRINT"TOOTH WIDTH W2 = ";INT(W2*100+.5)/100;"MM":PRINT
460 L4=1.15*S*W2/(π*P)
470 PRINT"YOKE DEPTH L4 = ";INT(L4*100+.5)/100;"MM":PRINT
480 REM HALF SLOT ANGLE IS C
490 C=π/S
500 A=.5*(D1*SIN(C)-W2)
510 B=.5*(D-D1)-L4
520 L5=(B*(1+SIN(C))+.2*A)/(1+.8*SIN(C))
530 PRINT"SLOT DEPTH L5 = ";INT(L5*100+.5)/100;"MM":PRINT
540 L6=(A+B*SIN(C))/(1+.8*SIN(C))
550 PRINT"SLOT RADIUS L6 = ";INT(L6*100+.5)/100;"MM":PRINT
560 L9=L5-(L6+L7+L8)
570 PRINT"THROAT TO RADIUS L9 = ";INT(L9*100+.5)/100;"MM":PRINT
580 W3=2*(L6-L9*SIN(C))/COS(C)
590 PRINT"MINIMUM SLOT WIDTH W3 = ";INT(W3*100+.5)/100;"MM"
600 A6=(L8*(W1+W3)+L9*(W3+2*L6)+π*L6↑2)/4
610 PRINT:PRINT"NET SLOT AREA A6 = ";INT(A6*100+.5)/100;"MM↑2"
READY.
```

Figure 11.4 Program for stator lamination design

behind the stator slot plus 20% of the radius of the slot bottom; this has been found to give a good average approximation to working flux densities.

Once again the microcomputer can be used with advantage in carrying out these calculations, as is shown in Figures 11.4 and 11.5. This program requires the input of four pieces of information, and can be used to provide a first look at a suitable lamination.

```
STATOR LAMINATION DESIGN
________________________

USE WITH DATA SHEET 19 PAGE 3

NUMBER OF POLES  4

STATOR OUTER DIAMETER (MM)  152

STATOR BORE (MM)  90

NUMBER OF STATOR SLOTS  36

SLOT OPENING W1 =  1.96 MM

THROAT DEPTH L8 =  1.13 MM

TOOTH WIDTH W2 =  3.96 MM

YOKE DEPTH L4 =  13.05 MM

SLOT DEPTH L5 =  18.6 MM

SLOT RADIUS L6 =  3.28 MM

THROAT TO RADIUS L9 =  13.45 MM

MINIMUM SLOT WIDTH W3 =  4.22 MM

NET SLOT AREA A6 =  46.41 MM↑2
```

Figure 11.5 Printout for stator lamination design

The design process may be iterated by inserting different values in this program, until a suitable lamination design has been achieved. By now the message will be clear, that much in the design of laminations is based on past experience from which empirical relationships have been derived. This reliance on past experience is not confined to small-power motor design; it occurs in almost every field of design. So when setting out to design both stator and rotor laminations, the designer must accept that the making and testing of a number of prototypes will be necessary before final design details can be determined, and it is here that the microcomputer really saves the designer valuable time, allowing concentration on decision making, while the machine handles all the boring and time-consuming calculations. Now let us look at rotor laminations.

Rotor laminations

The design of rotor laminations can also be handled sequentially, as shown in Figures 11.6, 11.7 and 11.8 (Data sheet 20 sheets 1, 2 and 3), and can be processed very

DATA SHEET

Source C V Starkey

Subject F H P Motors – Rotor Design **Date** April 1988

The major dimensions of the rotor stack have already been determined – see Data Sheet Number 19. The rotor outside diameter is fixed by the stator bore and the radial air gap, and the rotor stack length is usually made the same as for the stator.

Both stator and rotor laminations are normally punched from the same piece of lamination strip – typical thickness 0.45 mm. To complete the design of the rotor lamination, the number and size of slots and vent holes must be determined.

S = number of stator slots

S1 = number of rotor slots

For a satisfactory design, the following rules must be observed:-

S – S1 must not equal ± 1, ± 2, $\pm P$, $\pm(P \pm 1)$, $\pm(P \pm 2)$

S1 must not be equal to, divisible by, nor divisible into S

Make S1 divisible by P for a quiet motor – but with perhaps some 'cogging'

For quietness, make S1 differ from S by 20% or more.

For low reactance, make S1 larger than S

1. G = air gap = $0.127 + 0.0042\text{D1}/p^{1/2}$ mm
2. D2 = rotor outside diameter = D1 – 2G mm
3. W5 = rotor tooth width = 0.95 W2 S/S1 mm

   Rotor tooth-section area can be made 0.95 of stator tooth-section area, as the rotor teeth carry less flux. In addition, rotor tooth flux densities can safely be made somewhat higher than stator tooth flux densities.

For round slots:-

4. W4 = rotor slot opening, usually made 0.75 mm
5. W6 = rotor tooth tip, usually made 0.75 mm, but never less than 1.5 × lamination thickness, for tooling purposes.
6. d6 = rotor slot diameter = $\dfrac{\pi(\text{D2} - 2\ \text{W6}) - \text{S1 W5}}{(\text{S1} + \pi)}$ mm

For trapezoidal slots:-

7. W7 = rotor slot depth = 4 + 0.032D2 mm
8. θ = half rotor slot angle = $360^\circ/2\text{S1}$
9. d7 = rotor slot major diameter = $2\left(\dfrac{\sin\theta(0.5\ \text{D2} - \text{W6}) - 0.5\ \text{W5}}{\sin\theta + 1}\right)$ mm
10. W8 = rotor slot root radius = $\dfrac{(\text{W7} - 0.5\text{d7} - \text{W6} - \text{d7}/2\sin\theta)}{1 - 1/\sin\theta}$ mm
11. W9 = rotor yoke depth = 0.95 L4 mm

Data Sheet Number 20

Sheet 1 of 3

Figure 11.6 Data sheet for rotor lamination design

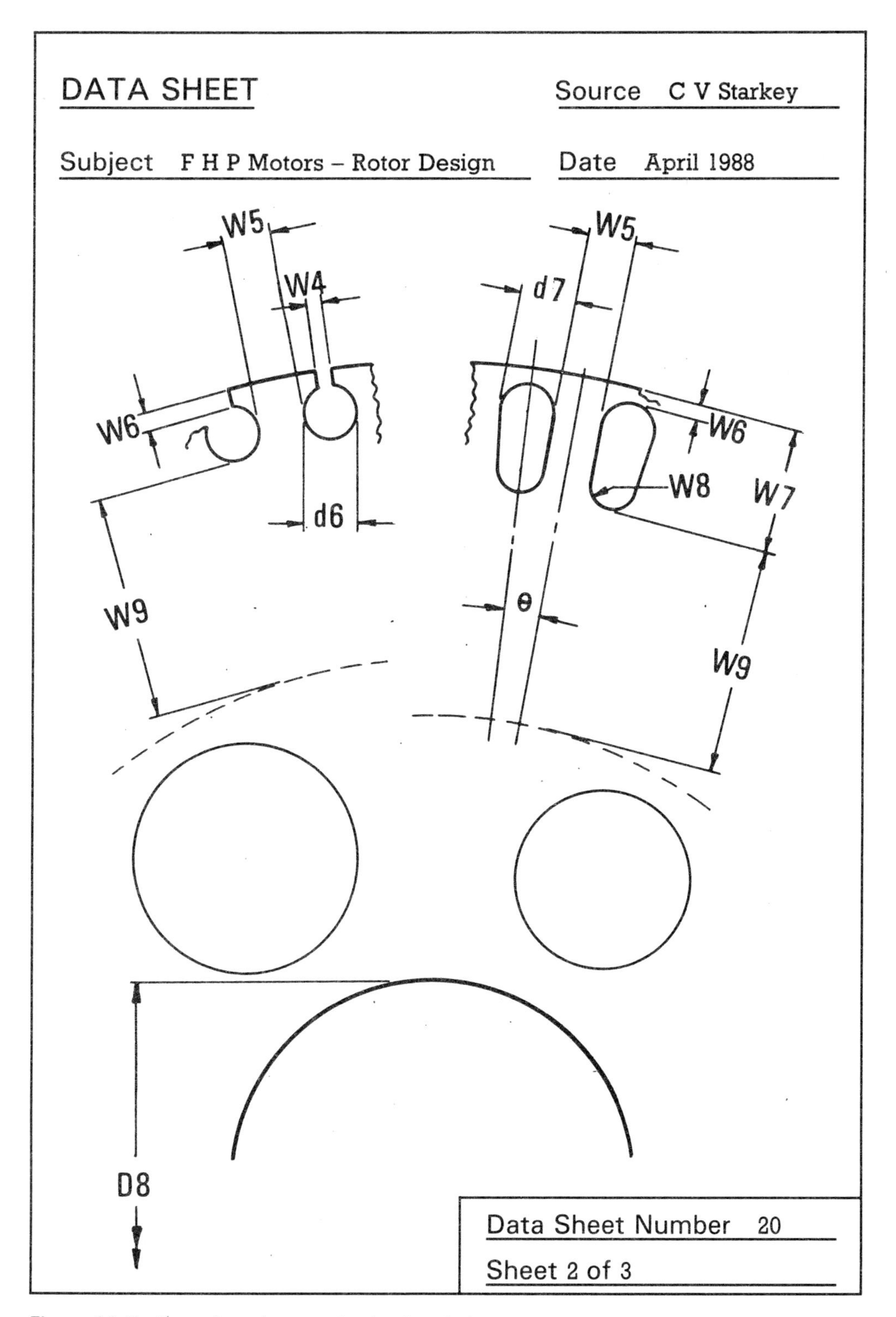

Figure 11.7 Data sheet for rotor lamination design

DATA SHEET

Source C V Starkey

Subject F H P Motors – Rotor Design

Date April 1988

Cross section of rotor end rings:-

Ar = cross sectional area of one end ring mm^2

Ab = cross sectional area of one rotor bar mm^2

Sl = number of rotor bars

P = number of poles

$$Ar = \frac{Ab\ Sl}{P\pi}\ mm^2$$

Data Sheet Number 20

Sheet 3 of 3

Figure 11.8 Data sheet for rotor lamination design

expeditiously by the computer program shown in Figure 11.9 giving the printout in Figure 11.10.

The cross-section of the rotor rings has been excluded from this program, as this is largely a matter of personal preference. Once the cross-sectional calculation has been completed manually, the actual geometrical shape of the rotor end ring is unimportant. It may be convenient to make the end ring deep in radial section and thin in axial length, or it may be shallow in radial depth and thick in axial length. Or it may be any combination between these two extremes, as the designer wishes.

```
5 PRINT CLS
100 PRINT"ROTOR LAMINATION DESIGN              "
110 PRINT"-----------------------      --------"
120 PRINT:PRINT"USE WITH DATA SHEETS 19 PAGE 3 & 20 PAGE 2":PRINT
130 INPUT"NUMBER OF POLES ";P:PRINT
140 INPUT"STATOR BORE (MM) ";D1:PRINT
150 INPUT"STATOR TOOTH WIDTH W2 (MM) ";W2:PRINT
200 INPUT"NUMBER OF STATOR SLOTS ";S:PRINT
220 PRINT"CHOOSE ROTOR SLOTS BETWEEN";INT(1.2*S);" AND";(2*S-1):PRINT
240 INPUT"NUMBER OF ROTOR SLOTS ";S1:PRINT
250 IFS1<INT(1.2*S)THEN220
260 IFS1>(2*S-1)THEN220
300 G=.127+(.0042*D1)/SQR(P)
310 D2=D1-2*G
320 W5=.95*W2*S/S1
330 W4=.75
340 W6=.75
350 D6=(π*(D2-2*W6)-(S1*W5))/(S1+π)
360 PRINT"ROTOR DIAMETER D2 = ";INT(D2*100+.5)/100;"MM":PRINT
400 PRINT"FOR ROUND SLOTS:":PRINT
410 PRINT"ROTOR SLOT OPENING W4 = ";W4;"MM":PRINT
420 PRINT"ROTOR TOOTH TIP W6 = ";W6;"MM":PRINT
430 PRINT"ROTOR SLOT DIAMETER D6 = ";INT(D6*100+.5)/100;"MM":PRINT
450 PRINT"PRESS ANY KEY TO CONTINUE"
460 GETA$:IFA$=""THEN460
500 PRINT"]":PRINT:PRINT
510 PRINT"FOR TRAPEZOIDAL SLOTS:":PRINT
520 INPUT"STATOR YOKE DEPTH L4 (MM) ";L4:PRINT
550 W7=4+.032*D2
560 REM F IS HALF ROTOR SLOT ANGLE
570 F=π/S1
580 D7=2*((SIN(F)*(.5*D2-W6)-.5*W5)/(SIN(F)+1))
590 W8=(W7-.5*D7-W6-(D7/(2*SIN(F))))/(1-1/SIN(F))
600 PRINT"ROTOR SLOT DEPTH W7 = ";INT(W7*100+.5)/100;"MM":PRINT
610 PRINT"ROTOR SLOT MAJOR DIAMETER D7 = ";INT(D7*100+.5)/100;"MM":PRINT
620 PRINT"ROTOR SLOT ROOT RADIUS W8 = ";INT(W8*100+.5)/100;"MM":PRINT
630 PRINT"ROTOR YOKE DEPTH W9 = ";INT((.95*L4)*100+.5)/100;"MM"
READY.
```

Figure 11.9 Program for rotor lamination design

```
ROTOR LAMINATION DESIGN
_______________________

USE WITH DATA SHEETS 19 PAGE 3 & 20 PAGE 2

NUMBER OF POLES  4

STATOR BORE (MM)  90

STATOR TOOTH WIDTH W2 (MM)  3.96

NUMBER OF STATOR SLOTS  36

CHOOSE ROTOR SLOTS BETWEEN 43  AND 71

NUMBER OF ROTOR SLOTS  48

ROTOR DIAMETER D2 =  89.37 MM

FOR ROUND SLOTS:

ROTOR SLOT OPENING W4 =  .75 MM

ROTOR TOOTH TIP W6 =  .75 MM

ROTOR SLOT DIAMETER D6 =  2.75 MM

PRESS ANY KEY TO CONTINUE

FOR TRAPEZOIDAL SLOTS:

STATOR YOKE DEPTH L4 (MM)  13.05

ROTOR SLOT DEPTH W7 =  6.86 MM

ROTOR SLOT MAJOR DIAMETER D7 =  2.75 MM

ROTOR SLOT ROOT RADIUS W8 =  1.14 MM

ROTOR YOKE DEPTH W9 =  12.4 MM
```

Figure 11.10 Printout for rotor lamination design

12 The Motor Windings

The design of windings for induction motors is not an exact science and involves some trial-and-error calculations. For this, if for no other, reason the microcomputer can be of immense use in handling iterative calculations, and avoiding the mistakes which so often occur when the designer becomes fatigued and boredom sets in. Figures 12.1, 12.2, 12.3 and 12.4 (Data sheet 19 sheets 5, 6, 7 and 8) show the sequential calculations for start and run windings suitable for a 4-pole machine using a 36-slot stator lamination.

The program in Figure 12.5 will handle the calculations for 2-, 4- and 6-pole machines with a 36-slot stator lamination. For other numbers of stator slots, it is necessary to alter the program to accommodate the changed requirement.

The two sizes of wire selected by the computer program, i.e. 0.92 mm for the run winding and 0.65 mm for the start winding, are not in the standard wire range. The nearest preferred sizes are 0.90 mm and 0.63 mm. The computer program could have been designed to search a database and recommend the two standard sizes, thus relieving the designer of the necessity of consulting a standard wire table. In fact, this technique will be used in the next section, which deals with the design of helical springs. Figure 12.7 shows an enlarged view of the start and run windings in the stator slot, together with the insulating slot liner and securing wedge.

Start winding switch

It will be remembered that a switch is used to disconnect the start winding from the circuit when the motor speed gets to 80% of synchronous speed on run-up. This switch is also used to reconnect the start winding if the motor speed should drop below 70% of synchronous speed through overloading.

There are many ways of detecting motor speed and using the resultant signal for control purposes. For example, a mark on the motor shaft may be sensed by a light-emitting-diode (LED) and the reflections counted by an integrated circuit on a microchip. By this method, the whole process is accomplished electronically, without any physical contact with moving parts of the motor. However, it does require the provision of a low voltage, direct current supply, and this adds expense to the method.

Alternatively, the motor shaft speed may be counted electrically by the insertion of a permanent magnet into the motor shaft which, in conjunction with an auxiliary lamination stack and winding assembly, will generate a voltage which varies with motor speed. This voltage could be used to indicate the 80% and 70% levels of synchronous speed and, through suitable control gear, to switch the start winding out or in as required.

DATA SHEET

Source C V Starkey

Subject F H P Motors – Stator Design

Date March 1988

Having determined the principal dimensions of the stator lamination and also the length of the stator stack for the power of motor required, the next step is to design the run winding, and then the start winding.

The most important consideration in the design of windings is to keep the flux densities in the various parts of the magnetic circuit at a reasonable level. For the purpose of estimating the run winding either the stator tooth flux density or the air gap flux density may be the controlling factor.

The following method considers the air gap flux density.

1. The design of the run winding is controlled by the back EMF formula:-

   Total stator conductors $C2 = EB\ 10^8/2.22\ \phi f$

   where: EB = back EMF = Supply voltage/1.1

   ϕ = air gap flux per pole = F A5

   F = flux density = 4650 to 6200 lines/cm^2

   A5 = area of one pole face = $\pi D1\ L/P\ cm^2$

2. The distributions of conductors in stator slots depends upon the number of slots per pole:-

   Slots per pole = Total stator slots/number of poles = S/P

   For a stator with 36 slots and 4 poles:-

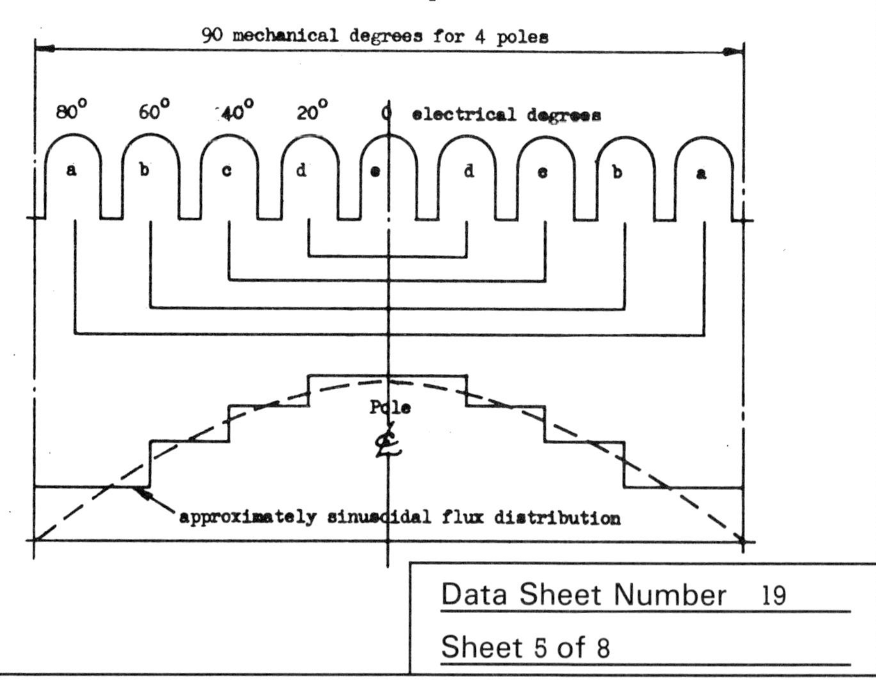

Data Sheet Number 19

Sheet 5 of 8

Figure 12.1 Data sheet for winding design

DATA SHEET

Source C V Starkey

Subject F H P Motors – Stator Design

Date March 1988

Coils are series wound and concentric and are laid in the slots as shown; coils which span 'a' slots are most effective, and coils which span 'd' slots are least effective. Effectiveness of coils is proportional to the electrical angle spanned; thus in the case shown on sheet 5 coil effectiveness is:-

for 'a' slots	$\sin 80^\circ =$	.985
" 'b' "	$\sin 60^\circ =$	.866
" 'c' "	$\sin 40^\circ =$	.643
" 'd' "	$\sin 20^\circ =$	.342
	Total =	2.836

and from this we can compute the fraction of the total conductors which are in each slot:-

fraction of total conductors in 'a' slots =	.985/2.836 =	.347
" " " " " 'b' " =	.866/2.836 =	.305
" " " " " 'c' " =	.643/2.836 =	.227
" " " " " 'd' " =	.342/2.836 =	.120

3. Winding distribution factor is derived from the combined effectiveness of all conductors in all slots:-

Effectiveness of conductors in 'a' slots = fraction of conductors in 'a' $\times \sin 80^\circ$
" " " " 'b' " = " " " " 'b' $\times \sin 60^\circ$
" " " " 'c' " = " " " " 'c' $\times \sin 40^\circ$
" " " " 'd' " = " " " " 'd' $\times \sin 20^\circ$

Winding distribution factor $C4 = (.347 \times .985) + (.305 \times .866) + (.227 \times .643) + (.120 \times .342)$

$$C4 = .793$$

As the total number of conductors in the stator are only .793 effective, we must increase that number to:-

$$\text{Total actual conductors } C1 = C2/C4$$

The actual number of conductors in an 'a' slot is $.347(C1/2P)$

4. Slot filling depends upon the net area available in the slot for conductors. The slot is lined with an insulating liner, wedges are forced into the slot to secure windings in position, conductors have an insulating sheath, and in slots which have run and start windings there may be insulating separators. Thus, only about half of the gross slot area is available for copper wire:-

$$\text{Net slot area } A6 = \frac{L8(W1 + W3) + L9(W3 + 2L6) + \pi L6^2}{4} \text{ mm}^2$$

Data Sheet Number 19

Sheet 6 of 8

Figure 12.2 Data sheet for winding design

DATA SHEET

Source C V Starkey

Subject F H P Motors – Stator Design

Date March 1988

To design the start winding, use 60% to 90% of the number of run winding conductors in a wire gauge giving half the cross sectional area, as a first approximation, and modify as necessary after testing.

The distribution of the total windings is such that if the number of stator slots divided by twice the number of poles does not equal a whole number, then either the run winding or the start winding must have an odd half coil per pole. As shown below, 36/8 = 4.5 and this results in 4 coils per pole for the run winding and $2\frac{1}{2}$ coils per pole for the start winding.

36 slot stator – Run Winding 4 coils per pole

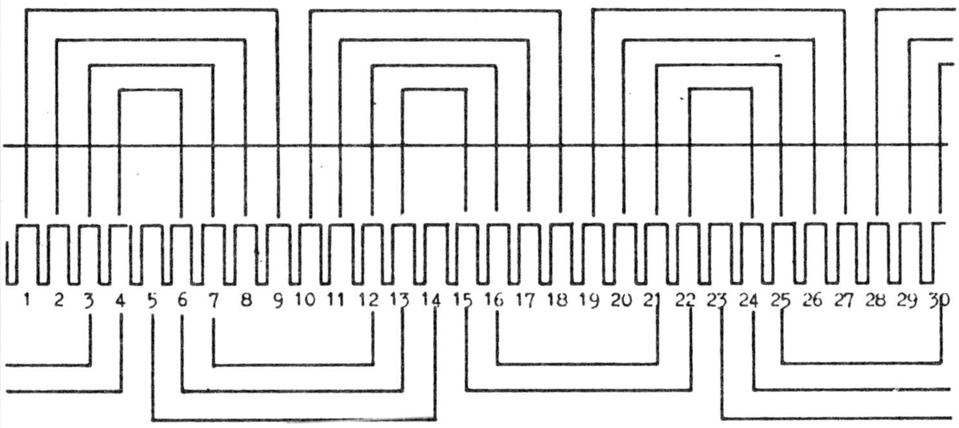

1 2 3 4 5 6 7 8 9 10 11 12 13 14 15 16 17 18 19 20 21 22 23 24 25 26 27 28 29 30

36 slot stator – Start Winding $2\frac{1}{2}$ coils per pole

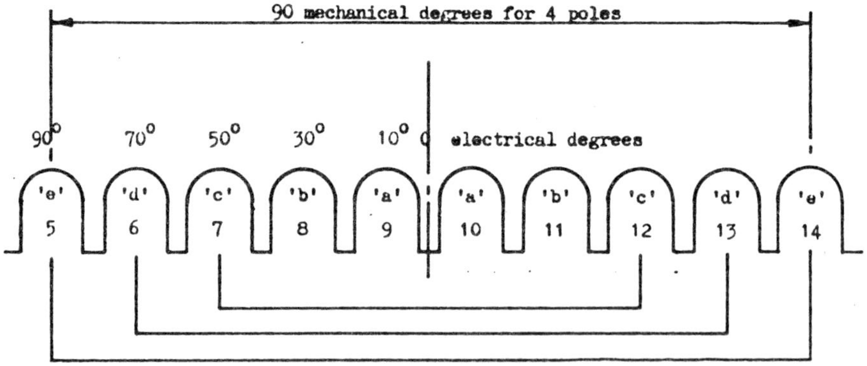

90 mechanical degrees for 4 poles

90° 70° 50° 30° 10° 0 electrical degrees

'e' 'd' 'c' 'b' 'a' 'a' 'b' 'c' 'd' 'e'

5 6 7 8 9 10 11 12 13 14

Data Sheet Number 19

Sheet 7 of 8

Figure 12.3 Data sheet for winding design

DATA SHEET — Source: C V Starkey

Subject: F H P Motors – Stator Design — Date: March 1988

Again, coils are series wound and concentric similar to the run winding; coils which span the 'e' slots are most effective, and coils which span 'c' slots are least effective. For approximately sinusoidal flux distribution, the coil effectiveness is:-

for 'e' slots	$\sin 90^\circ =$	1.000
" 'd' "	$\sin 70^\circ =$	.939
" 'c' "	$\sin 50^\circ =$	.766
	Total =	2.705

and from this, the fraction of total start winding conductors in each slot is:-

for 'e' slots	= 1.000/2.705 =	.370
" 'd' "	= .939/2.705 =	.347
" 'c' "	= .766/2.705 =	.283

Sizes of both the run and start winding conductors can now be calculated from the distribution of the combined windings. Wire sizes should be adjusted to the nearest preferred size from the R40 range from BS 3737:1964. (Data sheet number 4)

Assuming total start winding conductors = 0.6 total run winding conductors, ie 0.6C1/2P:-

		R W conductors		S W conductors
Conductors in 'a' slots	=	.347(C1/2P)	+	0
" " 'b' "	=	.305(C1/2P)	+	0
" " 'c' "	=	.227(C1/2P)	+	.283(0.6C1/2P)
" " 'd' "	=	.120(C1/2P)	+	.347(0.6C1/2P)
" " 'e' "	=	0	+	.370(0.6C1/2P)
Total conductors	=	C1/2P	+	0.6C1/2P

By inspection 'a' slots are the most densely packed, and from this we calculate the diameter of the run winding conductors:-

$$d4 = \sqrt{\frac{4(\text{net slot area A6})}{\pi(\text{no. of 'a' conductors})}}\ \text{mm}$$

For start winding conductors with half the cross sectional area of run winding conductors

$$d5 = \sqrt{\frac{d4^2}{2}}\ \text{mm}$$

Data Sheet Number 19

Sheet 8 of 8

Figure 12.4 Data sheet for winding design

```
5 PRINT CLS
100 PRINT"FHP WINDING DESIGN
110 PRINT"——————————————
120 PRINT:PRINT"FOR 36 SLOT STATOR LAMS ONLY"
130 PRINT"——————————————————————
140 PRINT:INPUT"STATOR BORE (MM) ";D1:PRINT
150 INPUT"STATOR LENGTH (MM) ";L:PRINT
160 INPUT"FLUX DENSITY (4650 TO 6200) ";F:PRINT
170 IFF<4650THEN160
180 IFF>6200THEN160
190 INPUT"SUPPLY VOLTAGE ";V:PRINT
200 INPUT"SUPPLY FREQUENCY (HZ) ";H:PRINT
210 INPUT"NUMBER OF POLES ";P:PRINT
220 IFP=2THEN270
230 IFP=4THEN600
240 IFP=6THEN800
250 GOTO210
260 REM ** TOTAL STATOR CONDUCTORS = C2**
270 A5=π*.01*D1*L/P
280 C2=((V/1.1)*100000000)/(2.22*F*A5*H):PRINT
290 C4=.852
300 C1=(C2/C4)/(2*P)
310 PRINT"□":PRINT
320 PRINT"          R W CONDUCTORS  S W CONDUCTORS
330 PRINT:PRINT"A SLOTS          ";INT(.184*C1);"              0"
340 PRINT:PRINT"B SLOTS          ";INT(.173*C1);"              0"
350 PRINT:PRINT"C SLOTS          ";INT(.152*C1);"              0"
355 PRINT
360 PRINT"D SLOTS        ";INT(.124*C1);"            ";INT(.066*.6*C1)
365 PRINT
370 PRINT"E SLOTS        ";INT(.093*C1);"           ";INT(.100*.6*C1)
375 PRINT
380 PRINT"F SLOTS        "INT(.061*C1);"           ";INT(.135*.6*C1)
390 PRINT:PRINT"G SLOTS          ";INT(.033*C1);"           ";INT(.165*.6*C1)
400 PRINT:PRINT"H SLOTS          0            ";INT(.187*.6*C1)
410 PRINT:PRINT"I SLOTS          0            ";INT(.199*.6*C1)
420 PRINT:INPUT"NET SLOT AREA A6 (MM↑2) ";A6
430 D4=SQR(4*A6/(π*INT(.184*C1)))
440 PRINT:PRINT"R W WIRE SIZE = ";INT(D4*100+.5)/100;"MM":PRINT
450 PRINT"S W WIRE SIZE = ";INT((SQR((D4↑2)/2))*100+.5)/100;"MM"
460 END
600 A5=π*.01*D1*L/P
610 C2=((V/1.1)*100000000)/(2.22*F*A5*H):PRINT
620 C4=.793
630 C1=(C2/C4)/(2*P)
640 PRINT"□":PRINT
650 PRINT"          R W CONDUCTORS  S W CONDUCTORS
660 PRINT:PRINT"A SLOTS          ";INT(.347*C1);"             0"
670 PRINT:PRINT"B SLOTS          ";INT(.305*C1);"             0"
680 PRINT:PRINT"C SLOTS          ";INT(.227*C1);"           ";INT(.283*.6*C1)
690 PRINT:PRINT"D SLOTS          ";INT(.120*C1);"           ";INT(.347*.6*C1)
700 PRINT:PRINT"E SLOTS          0";"           ";INT(.370*.6*C1):PRINT
710 INPUT"NET SLOT AREA A6 (MM↑2) ";A6
720 D4=SQR(4*A6/(π*INT(.347*C1)))
730 PRINT:PRINT"R W WIRE SIZE = ";INT(D4*100+.5)/100;"MM"
740 PRINT:PRINT"S W WIRE SIZE = ";INT((SQR((D4↑2)/2))*100+.5)/100;"MM"
750 END
800 A5=π*.01*D1*L/P
810 C2=((V/1.1)*100000000)/(2.22*F*A5*H):PRINT
820 C4=.856
830 C1=(C2/C4)/(2*P):PRINT"□"
840 PRINT"          R W CONDUCTORS  S W CONDUCTORS
850 PRINT:PRINT"A SLOTS          ";INT(.577*C1);"             0"
860 PRINT:PRINT"B SLOTS          ";INT(.422*C1);"           ";INT(.422*.6*C1)
870 PRINT:PRINT"C SLOTS          0            ";INT(.577*.6*C1)
880 PRINT:INPUT"NET SLOT AREA A6 (MM↑2) ";A6
```

```
 890 D4=SQR(4*A6/(π*INT(.675*C1)))
 900 PRINT:PRINT"R W WIRE SIZE = ";INT(D4*100+.5)/100;"MM":PRINT
 910 PRINT"S W WIRE SIZE = ";INT((SQR((D4↑2)/2))*100+.5)/100;"MM":END
READY.
```

Figure 12.5 Program for winding design

NOTE: in line 640, reverse field heart = clear screen

```
FHP WINDING DESIGN

FOR 36 SLOT STATOR LAMS ONLY

STATOR BORE (MM)?

 90

STATOR LENGTH (MM)?
 50

FLUX DENSITY (4650 TO 6200)?
 4650

SUPPLY VOLTAGE?
 250

SUPPLY FREQUENCY (Hz)?
 50

NUMBER OF POLES?
 4

                R W CONDUCTORS   S W CONDUCTORS

A SLOTS              68                0

B SLOTS              59                0

C SLOTS              44               33

D SLOTS              23               40

E SLOTS               0               43

NET SLOT AREA (MM↑2)?
 45

R W WIRE SIZE =  .92 MM

S W WIRE SIZE =  .65 MM
```

Figure 12.6 Printout for winding design

Mechanically, switching can be achieved using the old-fashioned, but highly reliable, principle of the governor, and this is the alternative we shall explore.

The switch consists of two separate subassemblies. The first of these is a static unit which is fixed inside the terminal end frame. The second is mounted on the motor shaft, between the rotor and the terminal end bearing, and rotates with the motor shaft.

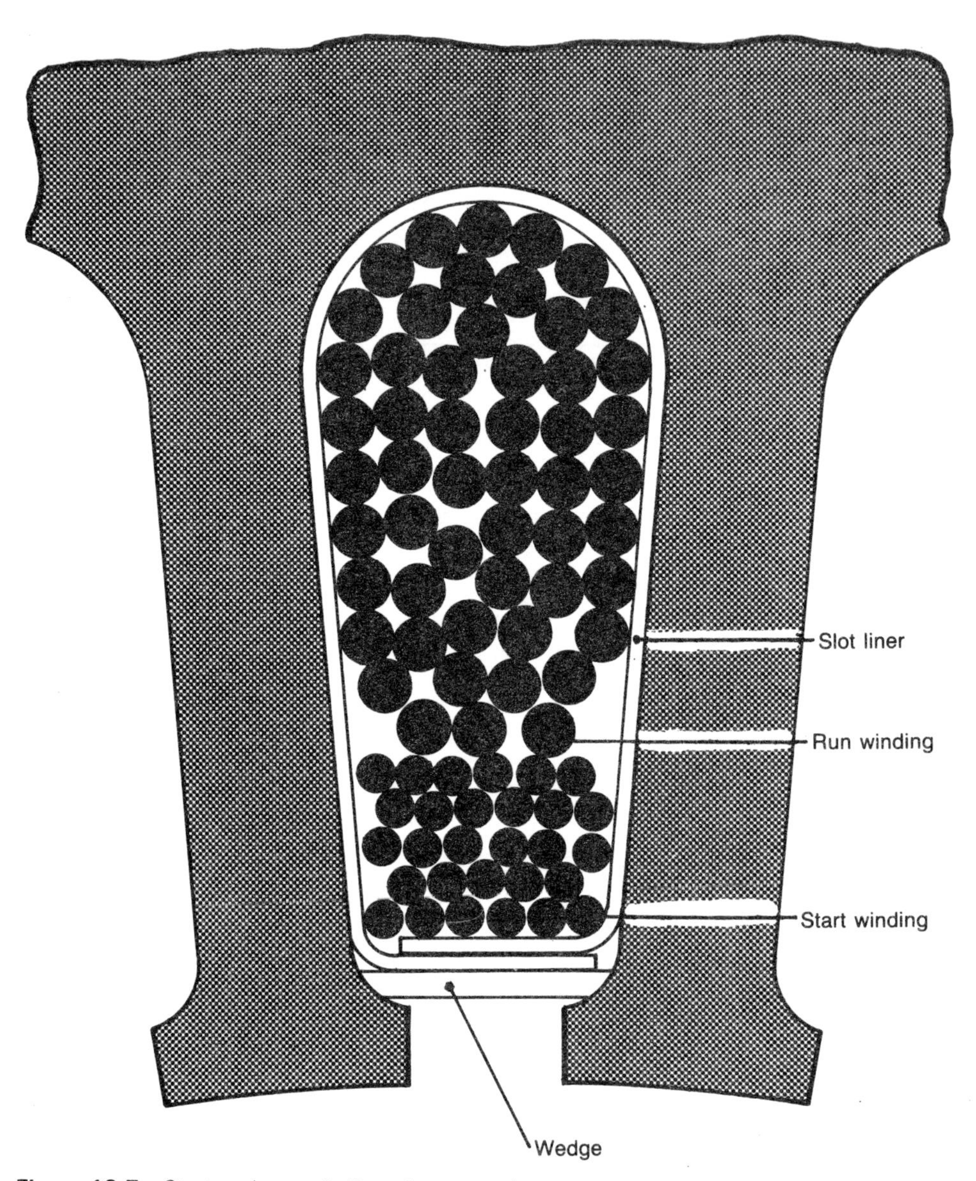

Figure 12.7 Start and run windings in stator slot

The static unit consists of a single-pole-single-throw switch. Its contact is carried on a forked member, which is pivoted and spring-loaded to give a **normally-open** condition.

The rotating unit comprises a supporting cradle in which two weights are pivoted. This can be seen immediately to the right of the rotor stack in Figure 6.3. The two pivoted weights engage a flanged bush which may slide along the motor shaft. The two weights are constrained by two helical extension springs, and when the motor is at rest, the two springs hold the two weights in toward the motor shaft. At the same time, the flanged bush is pushed along the shaft toward the terminal end, by a central projection from each of the two weights, where it bears on the forked member of the static unit, thus closing the switch in the start winding circuit.

On starting, the motor runs-up like a 2-phase machine and at 80% of synchronous speed the two weights fly outward around their pivots, sliding the flanged bush toward the rotor stack, and releasing the forked member of the switch. The static portion of the switch opens under its own spring-control and takes the start winding out of circuit. From this point on, the motor operates as a single-phase unit, for as long as the external load it has to satisfy does not become excessive. If the external load should become excessive, the motor will progressively slow down until it drops to 70% of synchronous speed. At this speed, the two weights snap inward toward the motor shaft, pushing the flanged bush toward the forked member, and the switch closes, bringing the start winding back into circuit. This immediately increases the torque available, and allows the motor to pick-up speed again.

In the switch sub-assemblies, the two components which are crucial to correct switch operation are the weights and the helical springs controlling them. The design of the weights is largely evolutionary, i.e. by trial-and-error based upon previous designs handled by the individual designer. Or it may be based on what competitive motor designers have achieved. It may even be designed from a suitable computer model, using perhaps 3-D modelling techniques at a CAD terminal. The important feature is to get the centre of gravity of the weight in the correct position vis-a-vis the centre about which the weight pivots. This is because it is necessary to build into the weight geometry a **differential** feature which, acting with the two springs, enables the governor to close at a speed which is lower than that at which it opens. This may involve considerable layout work on the drawing board coupled with repetitive calculations. Or it may involve repeated design changes at a CAD terminal. In either case it is beyond the scope of the present work, and we will accept a fait accompli, that a suitable weight has already been designed.

Figure 12.8 shows an exploded view of the principal components of the rotating sub-assembly of the switch, the governor portion. The pressed steel cradle is shown with the two pivots in-situ, the lower one anchoring the spring which controls the weight shown at the top of the sketch. The second spring and the other weight are deleted for clarity. The second weight revolves around the lower pivot and its controlling spring is anchored on the upper pivot. In each case, the spring hook is attached to a notched arm which extends sideways from the body of the weight.

Figure 12.9 shows the brass weight in both the closed and open positions relative to the motor shaft centreline. The centre-of-gravity of the brass weight is indicated in both positions, together with its dimensions from the motor shaft centreline and from the centre of the pivot. Also shown is the point at which the spring hook is attached to the weight, and the dimensions of this point relative to the pivot centre, in both the closed and open positions. By measuring the original of Figure 12.8, which was drawn to a scale of 10 to 1, it can be established that the extension of the spring, from the closed to the open position, is 2.9 mm.

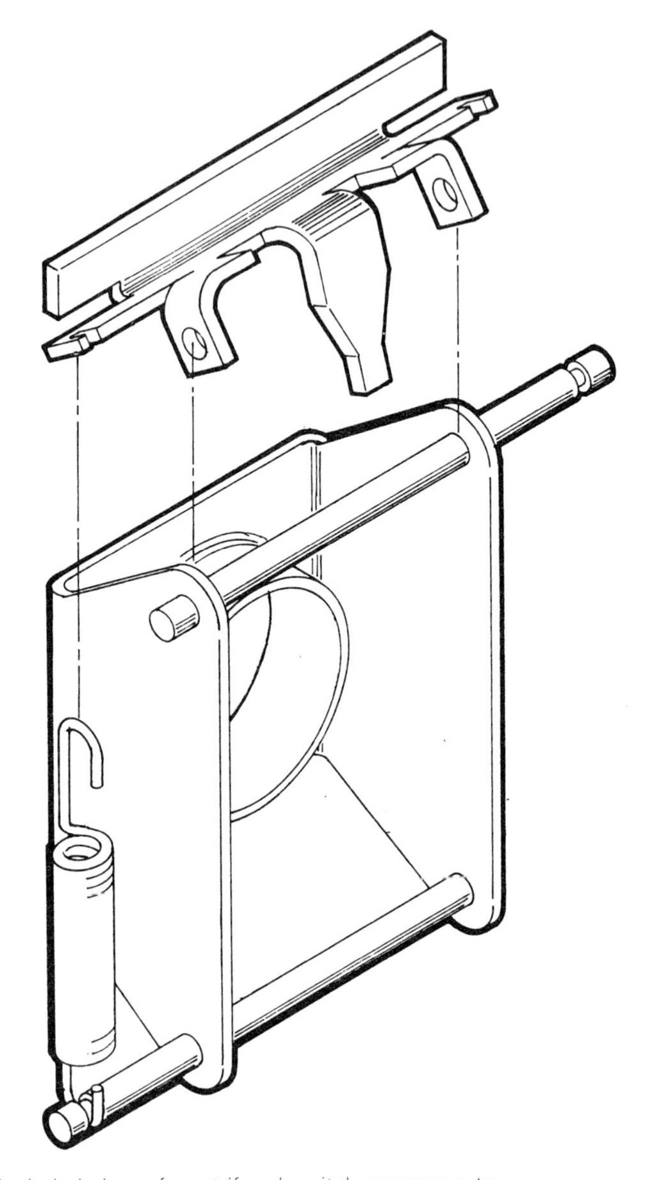

Figure 12.8 Exploded view of centrifugal switch components

Spring design

In order to design the extension spring for controlling the opening and closing of the centrifugal switch at the correct motor speeds, we must first establish the forces obtaining at the two positions of the brass weight. The mass of the brass weight has been calculated from a detailed drawing of the component, and it is 0.011 kg.

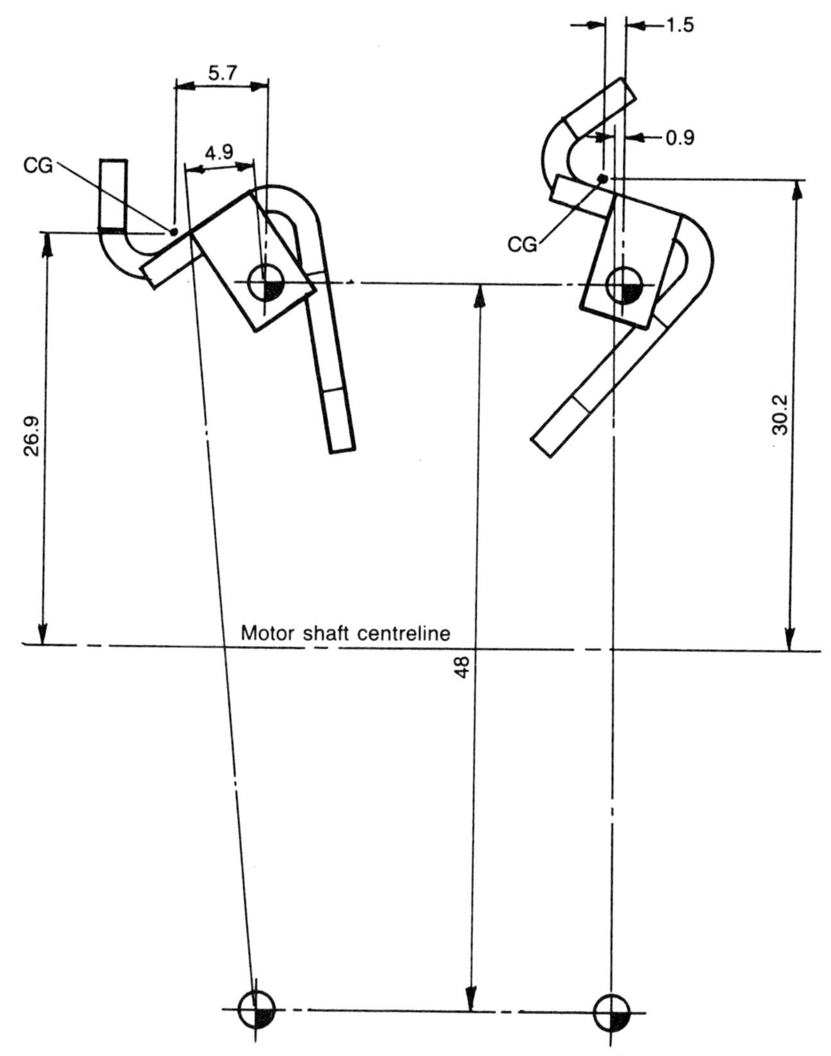

Figure 12.9 Spring-arm in closed and open positions

motor speed at opening = 0.8 × 1500 = 1200 rev/min. = 125.66 rad/s
motor speed at closing = 0.7 × 1500 = 1050 rev/min. = 109.95 rad/s

Centrifugal force acting on the brass weight is $M\omega^2r$ Newtons

where M is the mass of the brass weight in kg

and ω is the motor speed in rad/s

From Figure 12.9 r at opening speed is 26.9 mm

and r at closing speed is 30.2 mm

At 1200 rev/min.

centrifugal force $= 0.011 \times 125.66^2 \times 0.0269 = 4.673$ N
turning moment at pivot centre $= 4.673 \times 0.0057 = 0.0266$ Nm
spring force required for balance $= 0.0266/0.0049 = 5.43$ N

Note! The values 0.0057 and 0.0049 are taken from Figure 12.9 for the switch in the closed position.

At 1050 rev/min

centrifugal force $= 0.011 \times 109.95^2 \times 0.0302 = 4.016$ N
turning moment at pivot centre $= 4.016 \times 0.0015 = 0.006$ Nm
spring force required for balance $= 0.006/0.0009 = 6.69$ N

Note! The values 0.0015 and 0.0009 are taken from Figure 12.9 for the switch in the open position.

Thus, we have the two forces required of the spring to just balance the centrifugal forces applying at the closed and open positions. We also know that the extension of the spring between these two positions is 2.9 mm or 0.0029 m, and so we can now calculate the spring rate.

$$\text{spring rate } K = (6.69 - 5.43)/0.0029 = 434 \text{ N m}^{-1}$$

The spring will be made from piano wire with a safe working stress of 400 MN m^{-2}. The ratio of mean coil diameter to wire diameter, the spring index C, is normally selected from values between 3 and 12. A spring index $C = 3$ has the effect of raising the actual working stress in the wire by a factor of 1.58, while a value of $C = 12$ raises actual stress by a factor of 1.12. We will go for the middle of the range, $C = 8$ which raises actual stress by a factor of 1.18. See the following notes regarding Wahl's correction factor.

The general form of the spring formula is

$$\text{deflection } \Delta = 8WND^3/Gd^4 \quad \text{(see reference 16)}$$

where W is spring force at deflection Δ
N is the number of active coils
D is the mean coil diameter
G is the torsional modulus of rigidity
d is the wire diameter

From this we can calculate the number of active coils required.

$$N = 100(100d)^4/KD^3 = 100(100d)^4/434(8d)^3$$
$$N = 45003d$$

Wire diameter is calculated from the general form, taking into account Wahl's correction factor (4) which makes allowance for the residual stresses due to coiling, and which has the form

$$A = (4C - 1)/(4C - 4) + 0.615/C \quad \text{where } C = D/d \text{ the spring index}$$

In the present case, the spring index C has been set at a value of 8 which gives $A = 1.18$

$$\text{material stress} = 8AWD/\pi d^3 \text{ from which}$$

$$\text{wire diameter } d = \sqrt{8 \times 1.18 \times 6.69 \times 8/400\,000\,000\pi}$$

$$\text{which gives } d = 0.000\,634 \text{ m } (0.634 \text{ mm})$$

The nearest preferred size is 0.67 mm. So the number of active coils is

$$45003 \times 0.00067 = 30$$

$$\text{Maximum safe load on this spring is } 270(100d)^3/D = 15.15 \text{ N}$$

$$\text{Maximum safe deflection is } 0.027ND^2/d = 0.035 \text{ m (35 mm).}$$

Both of these values are well clear of the required performance of our spring, so it is unnecessary to check the worst case condition for maximum safe deflection.

Subject to satisfactory resonant frequency conditions, this spring will do the job. It is most important that the resonant frequency of the spring should not be near the running speed range of the motor, so as to avoid dangerous vibrations and possible spring failure.

```
5 PRINT CLS
100 PRINT"EXTENSION SPRING
110 PRINT"————————————————
120 PRINT:INPUT"FORCE AT INITIAL POSITION (N) ";P
130 PRINT:INPUT"FORCE AT FINAL POSITION (N) ";P1
140 PRINT:INPUT"EXTENSION - INITIAL TO FINAL (MM) ";B
150 PRINT:INPUT"SPRING INDEX (COIL/WIRE) ";C
160 REM**MAX MATERIAL STRESS = 400 MN/M↑2**
200 K=(P1-P)/B
210 N=100↑5/(1000*K*C↑3)
220 A=((4*C-1)/(4*C-4))+(.615/C)
230 D=SQR((8*A*P1*C)/(π*400))
240 PRINT
300 PRINT"WIRE DIAMETER = ";INT(D*100+.5)/100
310 FORI=1TO98:READV
320 IFV<DTHEN350
330 IFV>DTHEN360
350 NEXTI
360 PRINT:PRINT"PREFERRED WIRE SIZE = ";V;"MM"
370 N=INT((N*V/1000)+.5)
380 PRINT:PRINT"NUMBER OF COILS = ";N
390 PRINT:PRINT"SPRING RATE = ";INT(K*1000)/1000;"N/MM"
400 W=π↑2*(V/1000)↑2*C*(V/1000)*N*7840*9.81/4
410 F=93.6*SQR((K*1000)/W)
420 PRINT:PRINT"RESONANT FREQUENCY = ";INT(F);"CYCLES/MIN"
1010 DATA.036,.038,.040,.042,.045,.048,.050,.053,.056,.060
1020 DATA.063,.067,.071,.075,.080,.085,.090,.095,.100,.106
1030 DATA.112,.118,.125,.132,.140,.150,.160,.170,.180,.190
1040 DATA.200,.212,.224,.236,.250,.265,.280,.300,.315,.335
1050 DATA.355,.375,.400,.425,.450,.475,.500,.530,.560,.600
1060 DATA.630,.670,.710,.750,.800,.850,.900,.950,1.00
1070 DATA1.06,1.12,1.18,1.25,1.32,1.40,1.50,1.60,1.70,1.80
1080 DATA1.90,2.00,2.12,2.24,2.36,2.50,2.65,2.80,3.00,3.15
1090 DATA3.35,3.55,3.75,4.00,4.25,4.50,4.75,5.00,5.30,5.60
1100 DATA6.00,6.30,6.70,7.10,7.50,8.00,8.50,9.00,10.00
READY.
```

Figure 12.10 Program for extension spring

```
EXTENSION SPRING
________________

FORCE AT INITIAL POSITION (N)  5.43

FORCE AT FINAL POSITION (N)  6.69

EXTENSION - INITIAL TO FINAL (MM)  2.9

SPRING INDEX (COIL/WIRE)  8

WIRE DIAMETER =  .64

PREFERRED WIRE SIZE =  .67 MM

NUMBER OF COILS =  30

SPRING RATE =  .434 N/MM

RESONANT FREQUENCY =  16669 CYCLES/MIN
```

Figure 12.11 Printout for extension spring

resonant frequency $f = 93.6 \sqrt{K/H}$ cycles/min.
where K is the spring rate
and H is the weight of active coil material.

In the present case, the resonant frequency is 16 669 cycles per minute which is well clear of the motor speed range, and so this spring is satisfactory for our use. To summarise the specification for the extension spring ...

wire diameter	0.67 mm
mean coil diameter	5.36 mm
total number of close wound coils	30
type of spring cnds	long hook
load at extension 15.42 mm	6.69 N ± 0.5 N
spring rate	0.434 N mm^{-1}

Material: galvanised spring steel BS1408B range 3

Details of a suitable computer program and a printout for this spring are shown in Figures 12.10 and 12.11.

Finale

Thus far, we have looked at some of the more important design features of eight of the ten major assemblies of the motor ...

rotor and shaft assembly
terminal frame assembly
drive frame assembly
start and run windings
stator assembly
motor casing
bearing arrangement
start winding switch

and have made our decisions regarding the principal design aspects. We have put *some* of the flesh on the bare bones of our proposed new product. Much more still has to be done in terms of lower order decision making. For example ...

tolerances for the fit of the rotor on its shaft
geometry of rotor fan blades
details of end frames structures and lubrication
details of the cooling airflow system
assembly and fixing of the stator to the motor casing
sizes of, and stresses in, the motor tiebolts
geometrical and fixing details of the motor foot
geometrical tolerances of the switch sub-assemblies
details for inclusion on the motor nameplate
position and fixing of motor nameplate
details of corrosion-resistant finish for the motor assembly
product colour scheme

All of these, and many more, must be finalised before we have a fully detailed, viable working design of a motor. Perhaps the dedicated student may be disposed to progress further along this fascinating pathway, discovering en route just how absorbing and rewarding is the job of the engineering designer. Whether or not he or she chooses to go further, the major decisions have been made. The design process has been explored in some depth; an embryo product design exists as a base for further work.

In conclusion: this work has made no attempt to teach any technological material. But is has shown how technology is used by the designer as the design evolves from a sketchy base, through the accumulation of relevant disparate data, some of which may subsequently be modified by later decision making, to the final development of a viable product.

Design is essentially a learn-by-doing subject. Undergraduates with little or no industrial background cannot make sensible design decisions in the artificial environment of the classroom. So the guided design exercise with the small-power motor has been an attempt to plot a path showing how design decisions are made. En route, the student's choice of alternatives has intentionally been severely limited, in the interests of keeping the workload within the confines of the average design study period.

This has been an introduction to design procedures aided by a suite of simple BASIC programs, to integrate BASIC with ENGINEERING DESIGN and relieve the designer of much boring arithmetic. Hopefully, by doing so we have released time for the making of productive, cost-beneficial decisions for greater profitability.

From here on the Reader must go it alone, perhaps using this text as a useful reference source. Good luck in your future endeavours.

Bibliography and References

Bibliography

Chapter 1

Riggs J.L. and Inoue M.S., *Introduction to Operations Research and Management Science*, McGraw-Hill, NY, 1975.
Dieter G., *Engineering Design: a materials and processing approach*, McGraw-Hill, NY, 1983.
French M.J., *Conceptual Design for Engineers,* The Design Council, London, 1985.
Hubka V., *Principles of Engineering Design*, Butterworths, London, 1982.
Pahl G. and Beitz W., *Engineering Design*, The Design Council, London, 1984.

Chapter 2

Pritsker A.A.B. and Elliott Sigal C., *Management Decision Making: a network simulation approach*, Prentice-Hall, NJ, 1983.
Baker K.R., *Management Science: an introduction to the use of decision models*, Wiley, NY, 1985.
Forgionne G.A., *Quantitative Decision Making*, Wadsworth Inc., California, 1986.
Dickson D.N. Ed., *Using Logical Techniques for making Better Decisions*, Wiley, NY, 1983.

Chapter 3

Meredith D.D. et al, *Design and Planning of Engineering Systems*, Prentice-Hall, NJ, 1973
Riggs J.L., *Engineering Economics*, McGraw-Hill, NY, 1982.

Chapter 4

Jelen F.C. and Black J.H., *Cost and Optimization Engineering*, McGraw-Hill, NY, 1983.
Getz L., *Financial Management for the Design Professional*, Whitney Library of Design, Shoreditch, 1984.

Leech D.J. and Turner B.T., *Engineering Design for Profit*, Ellis Horwood, Chichester, 1985.

References

1 Alger J.R.M. and Hays C.V., *Creative Synthesis in Design*, Prentice-Hall, NJ, 1964.
2 Asimow M., *Introduction to Design*, Prentice-Hall, NJ, 1962.
3 Clugston R., *Estimating Manufacturing Costs*, Gower Press, London, 1971.
4 Duggan T.V., *Applied Engineering Design and Analysis*, Iliffe, London, 1970.
5 Evans R.K. and Hartley P.R., *Materials Selector and Design Guide*, McGraw-Hill, London, 1974.
6 Fitzgerald A.E. and Kingsley Jr. C., *Electric Machinery*, McGraw-Hill, NY, 1961.
7 Grant E.L., *Statistical Quantity Control*, McGraw-Hill, NY, 1946.
8 Gregory S.A., *Creativity and Innovation in Engineering*, Butterworths, London, 1972.
9 Kivenson G., *Durability and Reliability in Engineering Design*, Pitman, London, 1972.
10 Krick E.V., *An Introduction to Engineering and Engineering Design*, Wiley, NY, 1969.
11 Miles L.D., *Techniques of Value Analysis and Engineering*, McGraw-Hill, NY, 1961.
12 Starkey C.V., 'Electron Gun Assembly', *Engineer*, London, 1964.
13 Starkey C.V., 'Costs for Designers', *Engineering Designer*, London, 1976.
14 Svensson N.L., *Introduction to Engineering Design*, Pitman, London, 1976.
15 Veinott C.G., *Theory and Design of Small Induction Motors*, McGraw-Hill, NY, 1959.
16 The Spring Research Association Helical Springs Engineering Design Guides 08, Oxford University Press, 1974.
17 BS5000: Part 11, *Small Power Electric Motors and Generators*, British Standards Institution, London, 1973.
18 BS970, *Wrought Steels*, British Standards Institution, London, 1955.
19 BS2094, *Glossary of Terms Relating to Iron and Steel*, British Standards Institution, London, 1954.
20 BS4500, *Specification for ISO Fits and Limits*, British Standards Institution, London, 1969.
21 PD6470, *The Management of Design for Economic Manufacture*, British Standards Institution, London, 1981.

Index